SUN UP TO SUN DOWN

Shawn Buckley
Department of Mechanical Engineering
Massachusetts Institute of Technology

An Energy Learning Systems Book

McGRAW-HILL BOOK COMPANY

New York, St. Louis, San Francisco, Auckland, Bogota, Dusseldorf, Johannesburg, London, Madrid, Mexico, Montreal, New Delhi, Panama, Paris, Sao Paulo, Singapore, Sydney, Tokyo, and Toronto

TO HANK, MY MENTOR

Library of Congress Cataloging in Publication Data

Buckley, Shawn, date
Sun up to sun down.

Includes index.
1. Solar energy. I. Title.
TJ810.B82 621.47 79-13955
ISBN 0-07-008790-3

Copyright © 1979 by Shawn Buckley

All rights reserved. Printed in the United States of America. No part of this publication may be reproduced, stored in a retrieval system, or transmitted, in any form or by any means, electronic, mechanical, photocopying, recording, or otherwise, without the prior written permission of the publisher.

First edition, 1979

ISBN 0-07-008790-3

2345678910 MUMU 89876543210

Contents

Foreword

Our nation walks a tightrope today because of our addiction to foreign oil. The revolution in Iran and subsequent oil cutoffs demonstrated our vulnerability. Unlike the other energy crises of the 1970's, the most recent crisis won't go away. Shortages are going to get worse and the ramifications more serious. No matter how much we may wish for an energy panacea, there isn't one. The only cure is painful—withdrawal.

If you are feeling the pinch of the energy crisis now, you may be ready to help save this country from potential economic, political and environmental disaster. We live in a nation that demands an electric toothbrush and worries about nuclear waste disposal. We cannot have it both ways. Continuing on our present course invites the bureaucratic nightmare of gas rationing, a loosening of environmental protections, a serious loss of jobs, and increased environmental hazards from nuclear wastes, coal production and breeder reactors.

While there is no panacea, there are ways to wean ourselves from the oil well. Solar energy is an alternative—clean, renewable and safe. Solar offers the best means to get us through this century. It fosters energy conservation and self-reliance. Solar will create a new industry and thousands of new jobs. If we made solar and other renewable energy sources a national priority, we could meet 25 percent of our energy needs with these alternative sources by the year 2,000. Without a commitment to develop and commercialize solar, this nation's energy future is bleak.

Everybody likes solar energy. The success of "Sun Day" proved that the American people understand solar's benefits.

The problem is to transfer that positive public image into something people will buy and use. Making the basic concepts more easily accessible is one answer. Fortunately, Shawn Buckley, an MIT professor, is a technician with the ability to simplify solar technology for a general audience.

You and I don't have to wait another five years to use solar. Solar can meet some of our energy needs right now. Solar hot water and heating systems can be applied immediately. With the current oil price increases, solar is becoming more cost effective.

What I like about *Sun Up to Sun Down* "Understanding Solar Energy" is that it doesn't waste the reader's time trying to sell solar energy. It simply explains how solar technologies for hot water and heating work. The section on "solar economics" realistically explains its costs. The author uses descriptive analogies so the reader gets information without becoming bored or overwhelmed.

This book can move the reader from wondering about solar to becoming a solar consumer. I hope it moves thousands of Americans to make that choice. When I look at my two young daughters, I see energy as the greatest threat to their having healthful, productive lives. Our generation's failure to address this issue will hurt them far more than you or me. Those who read this book, advocate solar power, and install solar in their homes will be at the vanguard of an energy revolution. A revolution away from fossil fuels to renewable energy sources will make the United States more energy conscious and self-sufficient. The solar age will help give our children a comfortable and safe lifestyle.

Senator Paul E. Tsongas

Preface

I have written this book in an attempt to teach solar energy by analogy. Use of analogy in teaching technology is as nearly as old as technology itself. For example, when early electrical engineers sought to explain current flow in wires, they likened it to the flow of water through pipes. Many more people understood water flow in pipes; their knowledge could be extended to electricity by showing the analogy between water flow and electric current flow.

Even today, analogy makes learning easier. In my engineering courses at M.I.T., I teach that the behavior of systems are similar: not just fluid and thermal (i.e., heat) systems, but mechanical and electrical systems as well. When you learn how one "media" (as they are called) works, you can extend your knowledge to the other media very easily. All the media (fluid, thermal, electrical and mechanical) behave the same if the proper analogies are made. The analogy tool is used as a way to condense information. Instead of separately learning about each media, one is learned in detail and the knowledge is extended to the others. Often, on the first day of class I bring in a hot pad, a bicycle wheel, a cup of water, and an electrical circuit. Few students believe that these four "systems"—one from each media—behave the same way. Yet, by the end of the class I have shown how they are all closely related by analogy.

Analogy has been successfully used by others. Prof. Jay Forrester, of "Limits to Growth" fame, uses fluid analogies to convey his method of understanding social behavior. In a course I taught with Jay and several others at M.I.T., we showed how such diverse subjects as nuclear reactor behavior, the scrap metal market, the growth and decline of civilizations, and the national economy all have an analogous methodology. One of my contributions was, not surprisingly, the behavior of a solar heating system. Richard Feynman, Cal Tech's famous physicist, uses analogies quite successfully in his series of physics books. For example, electrostatic field theory is learned by analogy to stretched rubber membranes.

Knowing that thermal phenomena are difficult concepts to master, I felt learning by analogy could help transfer knowledge. I was left with a dilemma: which media should I use for the analogy. Had I been writing for electrical engineers I might have chosen an electrical media for analogy. Heat loss from a collector becomes an electrical resistor; heat storage becomes an electrical capacitor. In fact, my colleague, Richard Thornton of M.I.T.'s Electrical Engineering Department, teaches an excellent course on solar energy using just this analogy. But I felt that lay people are most familiar with fluid analogies and water flow is very easy to visualize. I find that imagining water squirting from a leaky rainbarrel forms an easier image than electrons leaking across capacitor plates.

A major problem with the analogy approach is that neither analogy may be familiar. To a person who doesn't understand either heat flow or fluid flow, the analogy approach may confuse more than it enlightens. I empathize with these people because there is no sugar coating for the pill. My approach is but one way to make the pill slide down easier. Learning is a difficult process and *any* aid to understanding is helpful. Of the lay people who initially found the fluid analogy confusing, most of those who stuck with it soon became comfortable with it. They claim now to understand *both* analogies quite well—the fluid and the thermal.

By using a fluid analogy to describe solar heating, I have been able to avoid mathematics. This may not be to everyone's liking; some readers are no doubt more comfortable with mathematical symbols and equations than they are with my somewhat contrived fluid analogies (the leaky pipe being the most notorious example). However, my experience has been that most lay people are not comfortable with mathematical language, including graphs and equations. That's not to say there are not difficult concepts presented in the book. Concepts such as stagnation temperature and collector efficiency are difficult to understand whether math or analogies are used. I've tried to avoid as much as possible the very technical aspects of solar engineering. I've slid past a great many details in hope of elucidating the major points. For example, I'm sure my portrayal of radiation heat flow caused many winces by my colleagues. Nevertheless, I feel the book is fairly complete since most aspects, including some rather technical material, are well covered.

Although I have been rather glib in presenting fluid analogies to thermal phenomena without resorting to mathematics, there is a very analytically precise methodology behind my presentation. Each concept is backed by rigorous mathematic analogies which have been taught to M.I.T. students for twenty years. I cannot take credit for developing the methodology although I am proud to have worked with those who did. Simply by adding appropriate equations for concepts I have discussed in prose, this book could become a college engineering text.

Cambridge, Mass.
April 1979

Acknowledgments

I would like to thank the following people who helped me produce this book:

Robert Entwistle: Bob, more than any other person, was responsible for this book being written. As it evolved from first a text and then into a more general book, it was Bob who was always prodding me to finish those last few chapters. His company, Energy Learning Systems, is co-publishing the book with McGraw-Hill. His production specialists have transformed my clumsy wording and crude figures into a polished tome.

Aldo Spadoni: Aldo, a mechanical engineering student at M.I.T. was the ideal person for the illustrations. Not only is he an excellent artist, but he understood the technical aspects of what I was trying to convey. We struggled for weeks to get the proper illustration style: clear but not intimidating.

Maureen Sheehan: Maureen had to have the patience of Job to type and retype and retype again each of many drafts; all with nary a complaint.

Gabriel Berriz: for his excellent printing that accompanies the illustrations.

Richard Thornton: who gave me invaluable advice on aspects of solar energy with which I was not familiar.

Patricia Dinneen: Last, but far from least, Patty gave me encouragement when I was down, advice when I came to stumbling blocks, and most important, provided a supportive environment.

The concept of consistent analogies between thermal fluid, electrical, and mechanical mediums has been pioneered by my colleagues, notably Hank Paynter, Art Murphy, Lowen Schearer and Herb Richardson. I am indebted to them for laying out the fundamental structure which I used in this book.

INTRODUCTION AND ECONOMICS 1

The sun has been giving energy to the earth for millions of years. Sunlight shining down on plants eons ago was the major cause of the coal, oil, and natural gas we use as fuels today. These fuels are called *fossil fuels* because they are the products of fossils of dead plants.

Fossil Fuels: Ancient Solar Energy

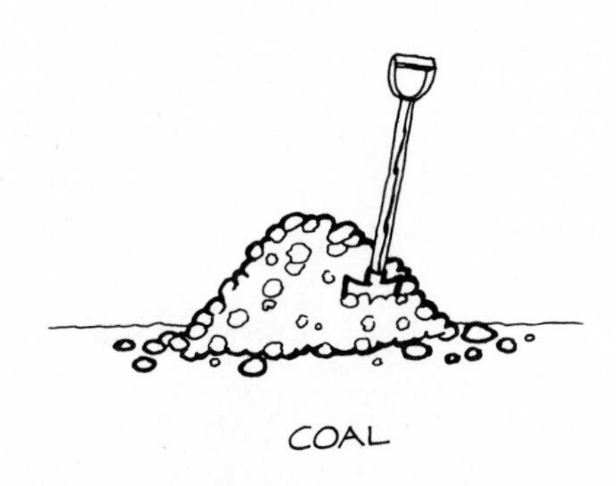

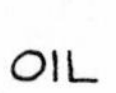

A similar source of solar energy used today is called *biomass;* this term refers to the use of the sun's energy to grow plants which can be used for fuels and other purposes. For example, wood used in fireplaces is a form of biomass solar energy. Researchers are working on other ways to grow plants for use as energy. Seaweed is being grown and harvested for methane production (a gas–like natural gas), and trees are being grown for use in power plants to generate electricity.

Wood:
Indirect Solar Energy

The sun also has a major impact on wind and ocean currents. Different parts of the earth's surface are heated differently by the sun, causing circulating currents in both the oceans and the atmosphere. Ocean currents are rivers of cold or warm water which can be tapped to run power plants; someday, these currents may be used to produce electricity. Winds have been used for centuries to power pumps and to grind grain.

Wind:
Indirect Solar Energy

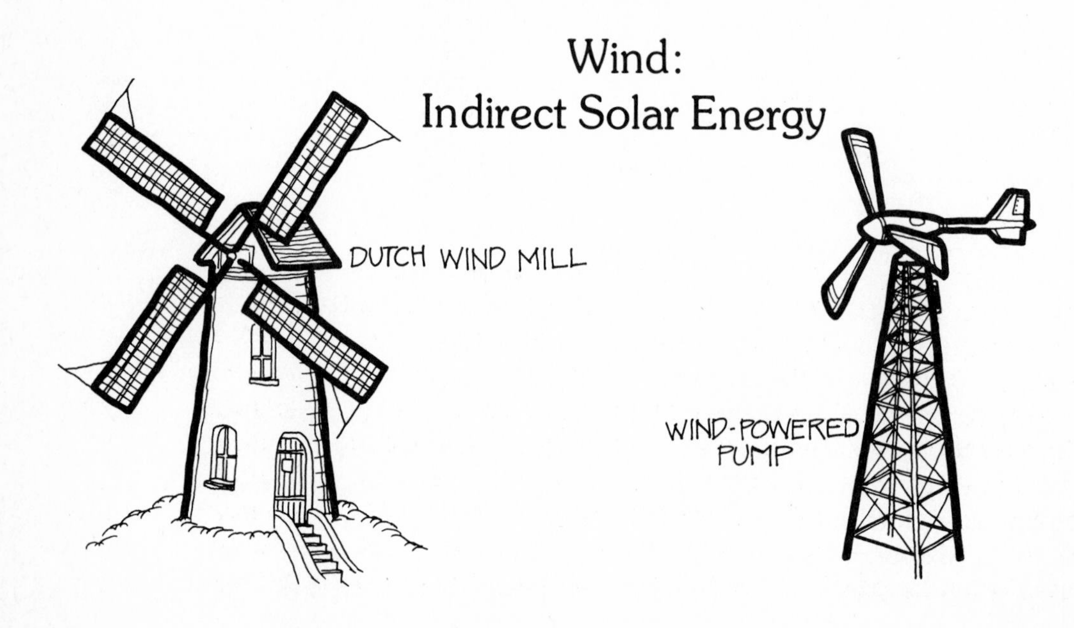

The sun's radiation may also be used *directly* to provide either electricity or heat; it is used directly as it comes raining down through the atmosphere rather than *indirectly* by making plants grow or causing wind or ocean currents.

Solar cells convert sunlight directly to electricity. These cells are usually made of silicon (a mineral used in the transistors which run radios and pocket calculators). Many of the space satellites circling the earth today are powered by solar cells. Since solar cells are very expensive in terms of the amount of electricity they provide, their use to date has been mostly to provide power for remote weather stations where the cost of bringing in power lines would be prohibitive; another application has been in powering remote railroad-crossing lights and ocean buoys.

Solar Cells: Direct Solar Energy

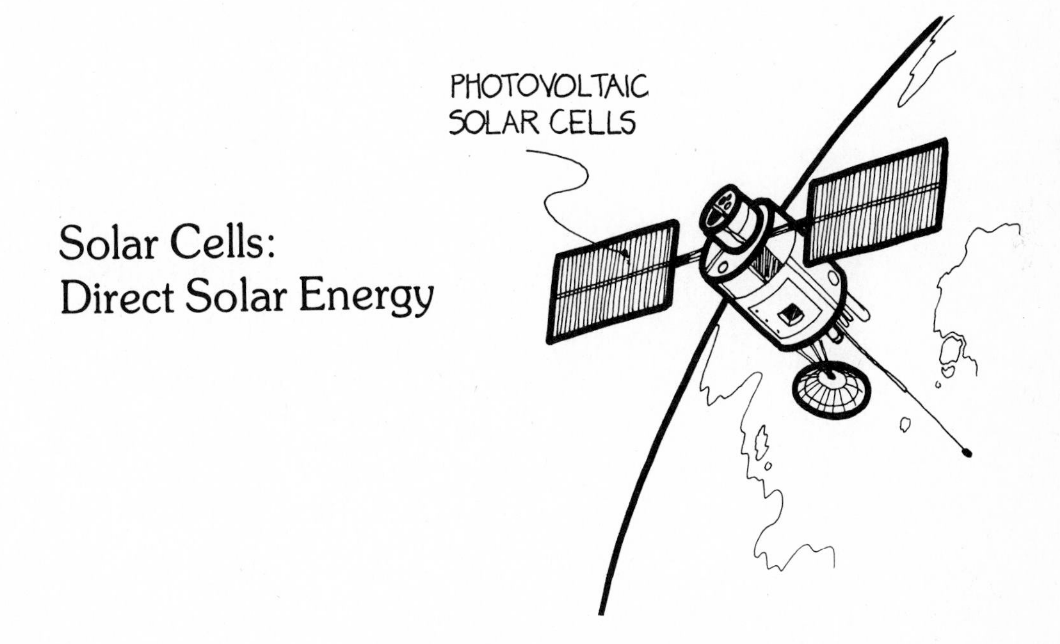

Heating is the other example of direct use of the sun. You've probably experienced the heating effects of solar energy when walking barefoot on a tarred road on a hot, sunny day. The heat produced by solar energy can be used to power turbines to generate electricity, or it can be used to heat buildings or water. One system now being developed to generate electricity uses hundreds of mirrors which focus the sun's rays on a central tower; a boiler on the tower gets very hot and produces steam, driving a turbine that generates electricity. Although the sun's position changes from morning to evening, the mirrors move to

Solar Cells Used on Earth

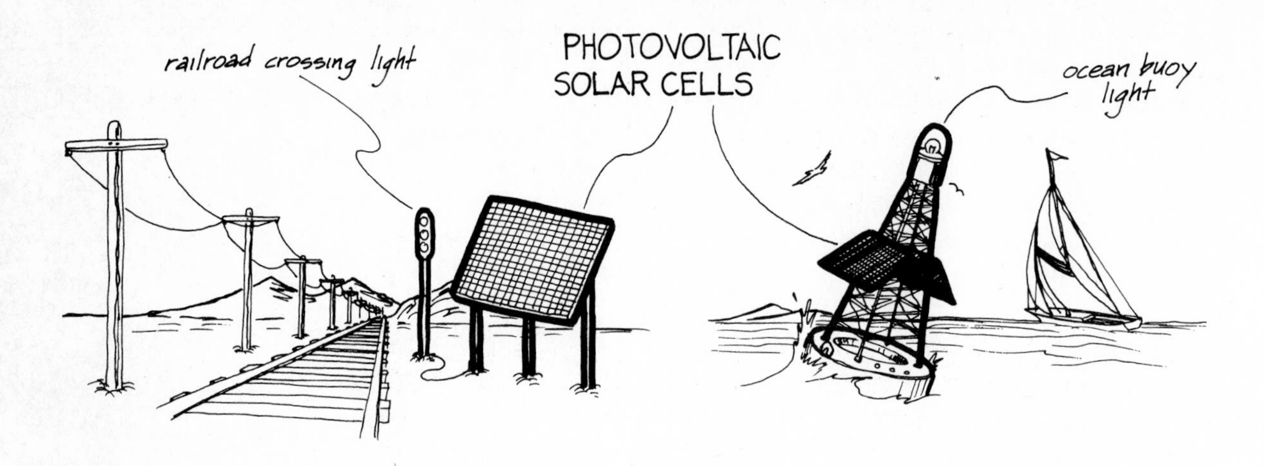

let the sun's reflection focus on the tower. Systems like these are called *tracking* because they follow the sun from sunrise to sundown.

Even if a solar energy system *isn't* designed to follow the sun, the sun's heat can still be captured. Solar heating systems which aren't of the tracking kind are called *flat-plate* systems. In comparison to tracking systems, they are not very sophisticated, but almost all of the solar energy systems in use today are of the flat-plate variety. Over a million flat-plate systems are

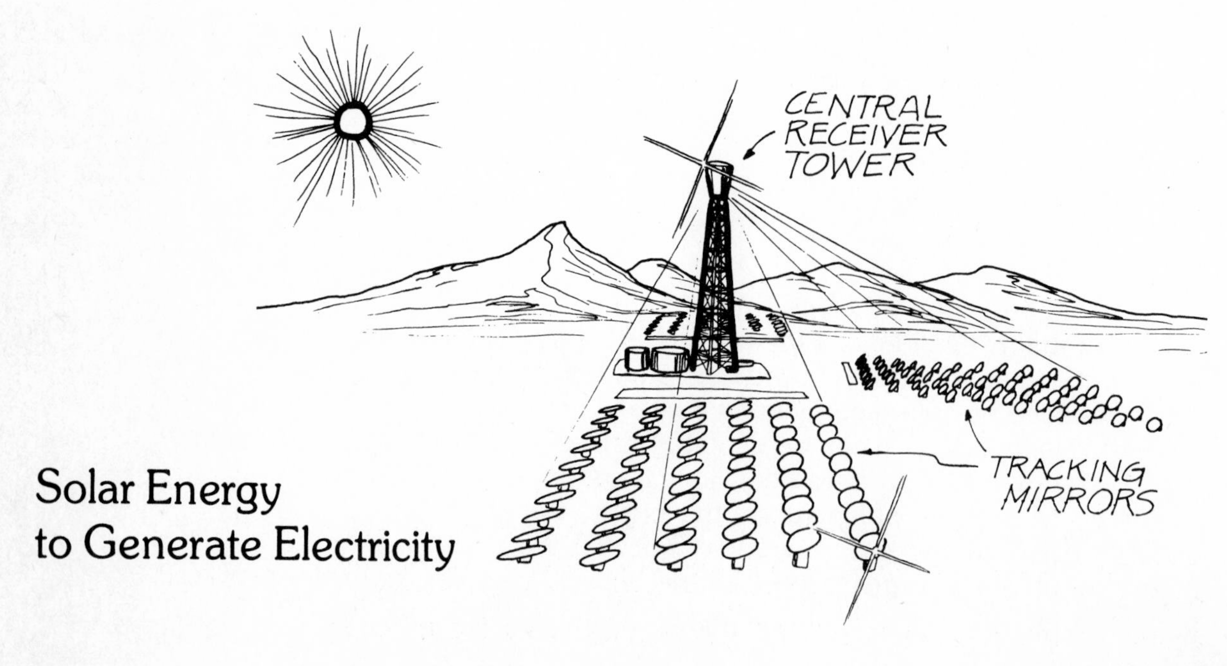

Solar Energy
to Generate Electricity

now in operation, mostly in countries with hot, sunny climates such as Israel and Australia. The most common use for such systems is to heat water, but it is also practical to use them to warm a house in winter. Since flat-plate systems are the simplest and most widely used methods of using solar energy today, we will investigate them in detail. But first it is important to understand why solar energy has just recently begun to be a practical means of providing direct heating.

If sunlight is free, why hasn't solar energy been used before to heat houses and produce electricity? The following example will help you understand why. Let's suppose you owned a gold mine that contained only very low-grade gold ore. You would have to do a lot of digging before you got even a little gold. If you bought some expensive mining equipment, you could process much more of the low-grade ore and get more gold. Thus, even though the gold itself is free, it would be very costly to get very much of it out of the mine. You would have to balance the cost of the mining equipment against how much money you'd get when you sold your gold. If the gold were very valuable and the mining equipment relatively cheap, it might be worthwhile to dig up the gold. But if the equipment were expensive and the gold not very valuable, you'd be better off to leave it in the ground. Many gold mines in Utah that have only low-grade ore were closed down years ago; some have recently reopened because the price of gold has increased enough to make mining it worthwhile.

Solar energy is like the low-grade ore. The sun's rays must be "mined," or *collected*, and then transformed into useful heat or electricity before they are worth anything. A solar energy system helps you get "free" solar energy, just as the mining equipment helps you get "free" gold. But, like mining equipment, a solar energy system can be very expensive—perhaps more than the sun's energy is worth.

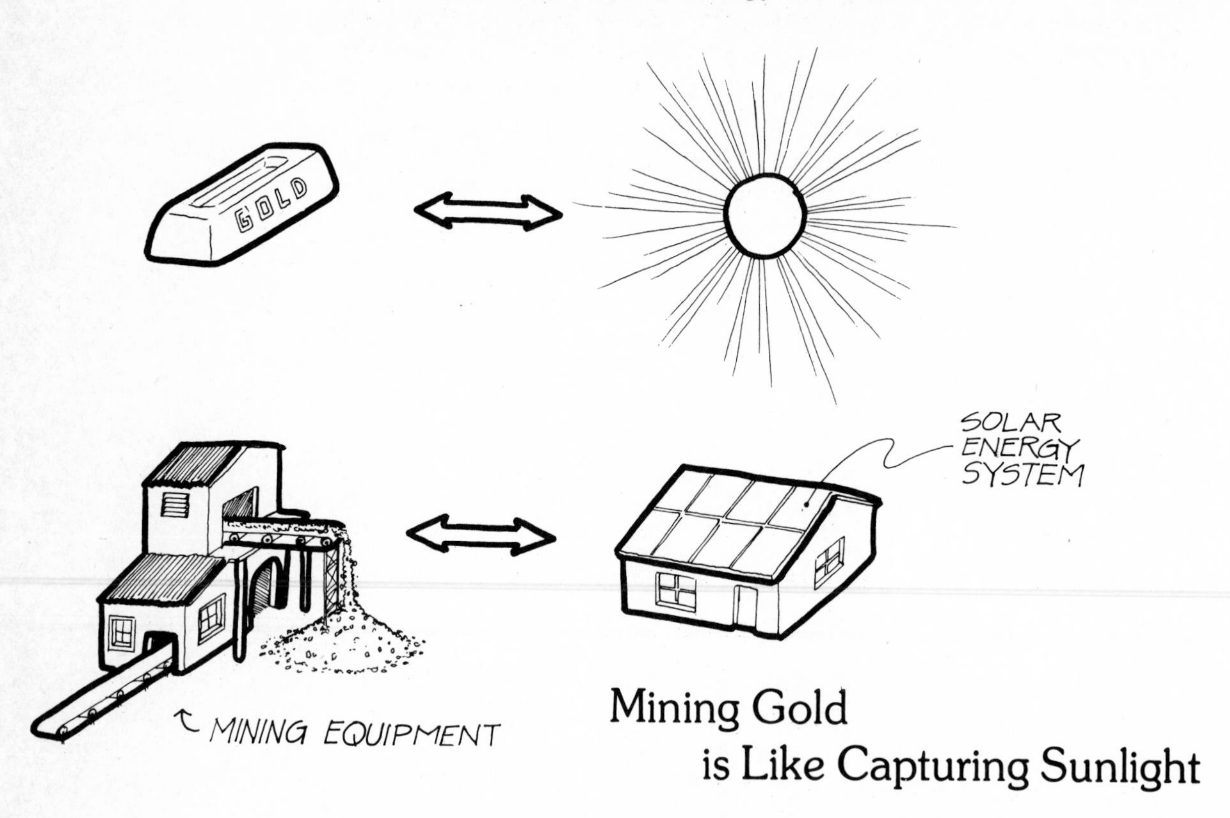

Mining Gold is Like Capturing Sunlight

If conventional energy is expensive—that is, if the cost of heating by using fossil fuels and electricity goes up—then it might be worth the expense to install a solar energy system. But if these forms of energy are cheap, then the cost of installing a solar energy system would probably be too expensive. Simply put, the cost of the solar energy would come to more than the value of the energy it could collect from the sun. Just as many gold mines in Utah were shut down when it became too costly to mine their gold, many solar hot-water heaters used in Florida and California during the 1950s were shut down when the cost of electricity was so cheap. Recently the cost of heating by gas, oil, and electricity has risen so much that solar energy systems are once again worthwhile, much as the Utah gold mines have once again become profitable because of the rise in the price of gold.

A solar energy system can be thought of as "mining" sunlight. If mining equipment is simple and rugged, it will give many years of service before it must be replaced. But mining equipment can't be too expensive or it will make mining low-grade ore unprofitable. Like mining equipment, solar energy systems must be simple, rugged, and low in cost so that they will give many years of service without making the sunlight that they "mine" unprofitable. This is why the *flat-plate collector* solar system is being used so much today in countries such as Japan, Israel, and Australia: it is a very simple, rugged system whose cost is low enough to make the use of solar energy worthwhile.

HEAT, TEMPERATURE, AND HEAT FLOW 2

The concepts of *heat* and *temperature* must be understood before a solar energy system can be understood. Often these two terms are misunderstood and, even worse, used interchangeably. However, these words have quite precise meanings.

Heat and temperature differ in important ways. First, temperature can be measured directly with a thermometer, but heat must usually be measured indirectly. How hot something is, is a measure of its temperature, but how much heat it contains is not so easily determined. In fact, the heat contained in an object usually depends not only on what it's temperature is but also on other factors, such as what it is made of and how much of it there is.

Heat and temperature differ in another important way: heat costs money. When you heat your house for the winter, you buy heat from the local heat dealer—from the oil man or the gas or electric company, depending on how your house is heated. Whatever form your heat comes in, what is sent to your house costs money. The more heat that is sent, the higher the heating bills are. But you don't buy temperature. Temperature indicates how hot your house is, not how big your heating bills are. When considering solar energy systems in the pages that follow, always remember that what is wanted is *heat*, not temperature. The only way to save money by using a solar energy system is to have the sun pay for part of your heating bill—to provide you with heat.

A simple analogy to the heat and temperature of an object is the volume and depth of water in a cup. How deep the water is does not measure the cup's volume. Although the volume and the depth are related—generally, the deeper the water in the cup is, the more volume of water there is in it—depth is a different measure from volume. Also, the depth of the water can be mea-

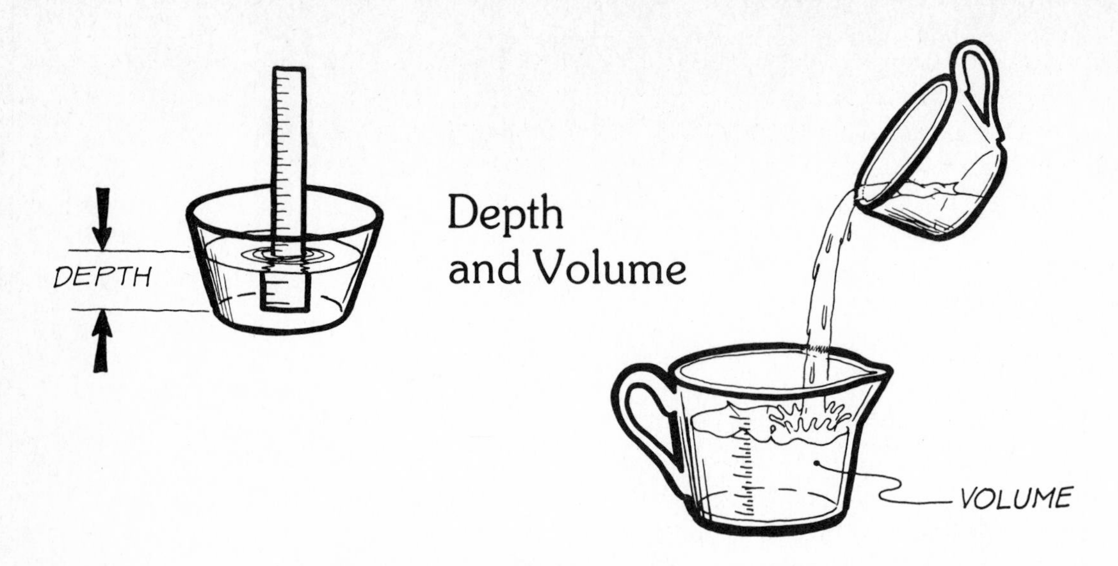

sured with a ruler, but a measuring cup must be used to ascertain the volume. According to our analogy, depth and temperature are easily measured, so they are like each other; volume and heat are measured only indirectly, so they are like each other too.

DEPTH ⟺ TEMPERATURE

VOLUME ⟺ HEAT

To see in more detail how depth is related to temperature, and volume to heat, let's consider a few simple cases. First, think of two different-sized cans that have the same depth of water in them. Though the depth of water in both cans is the same, the bigger can holds more volume. Similarly, if two pans

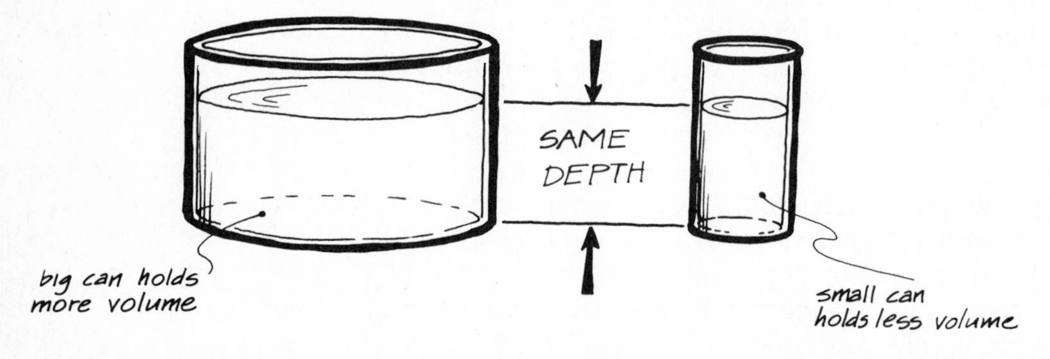

Same Depth but Different Volume

of dirt are heated in an oven for several hours, both will have the same temperature. But even though the temperature of both pans is the same, the bigger pan holds more heat. Just as the depth can be the same while the volume is different, so can the temperature be the same while the heat is different.

Same Temperature but Different Heat

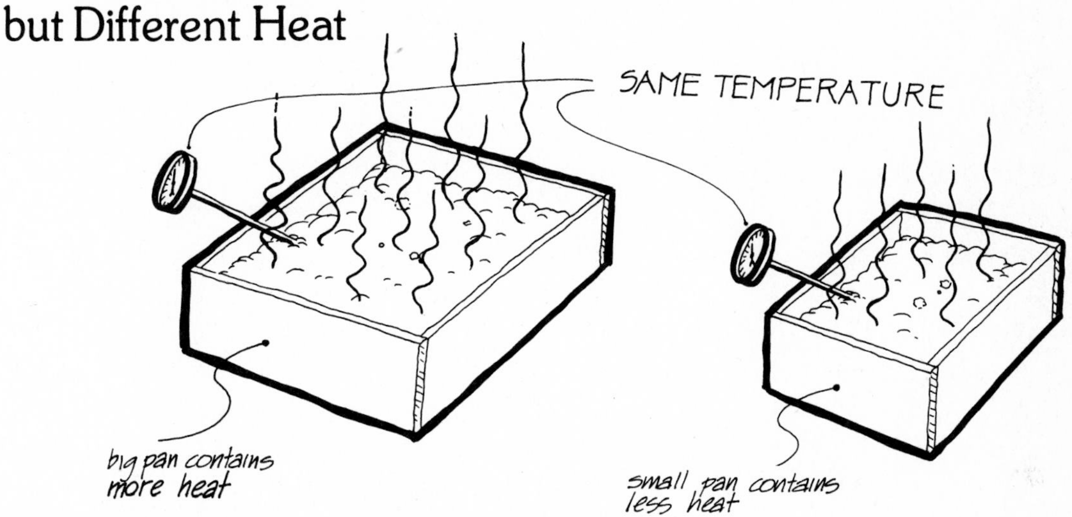

Second, let's consider two objects in which the heat is the same yet the temperatures differ. Two warm bricks could contain the same amount of heat as one hot brick; though the hot brick naturally has a higher temperature than the warm ones.

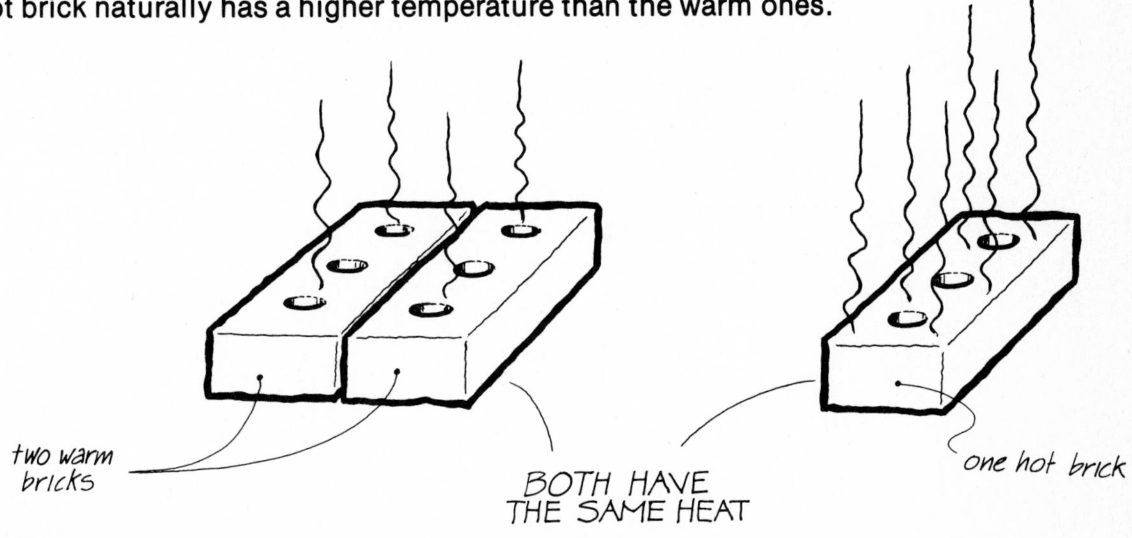

Same Heat but Different Temperatures

In the same way, two cans can hold the same volume but have different depths.

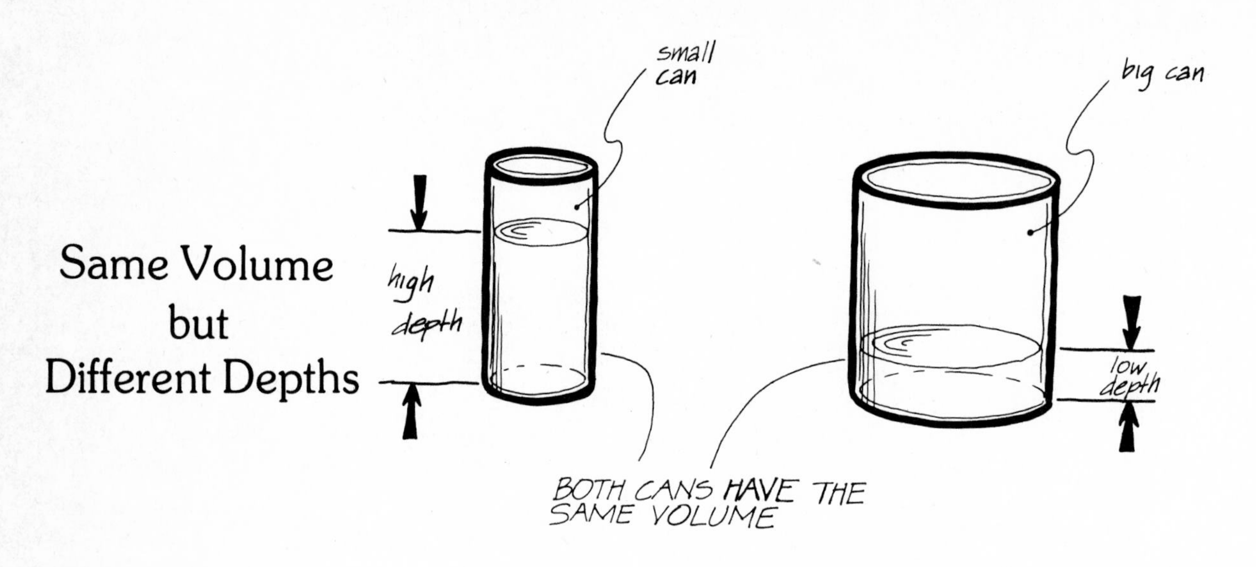

The analogy is helpful in another way. Both volume and heat *flow*—both can move from place to place. When you pour water out of a pitcher into a glass, volume flows from the pitcher to the glass; when you open a door on a cold day, heat flows from inside the house to the outdoors.

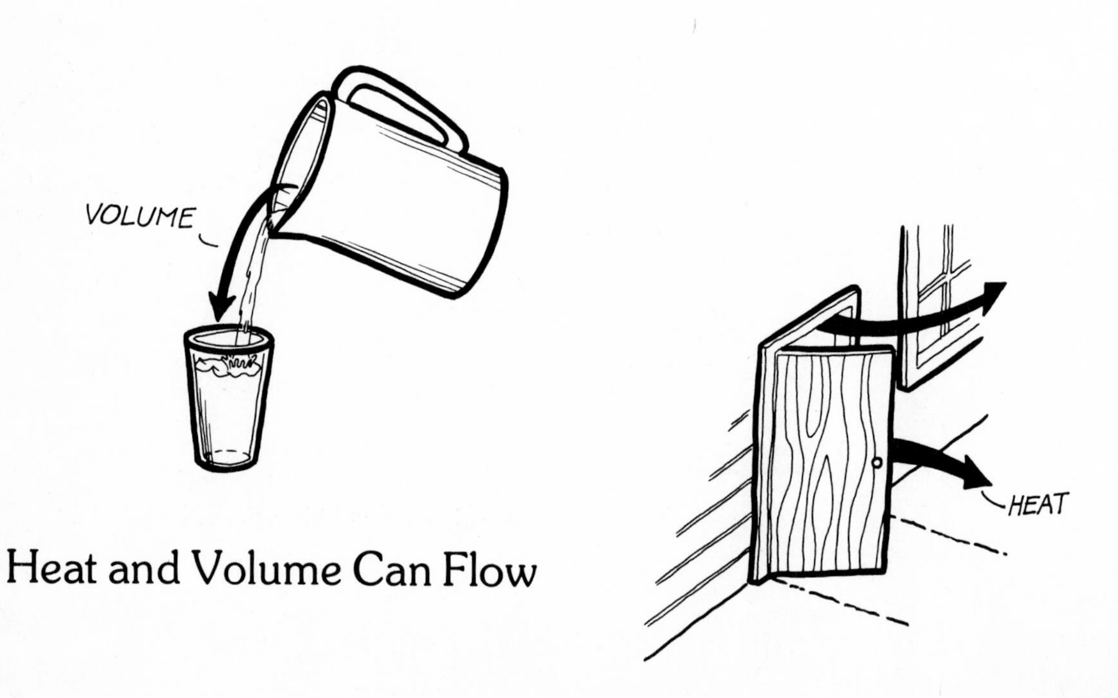

Why does heat flow? Heat flows from one place to another because the temperature of the two places is different. A hot brick loses heat to a cool room. The *temperature difference*—the brick's temperature minus the room's temperature—drives the heat from the brick. Heat leaks from the brick until the temperature difference is gone. No more heat flows from the brick when it becomes as cool as the room it is in.

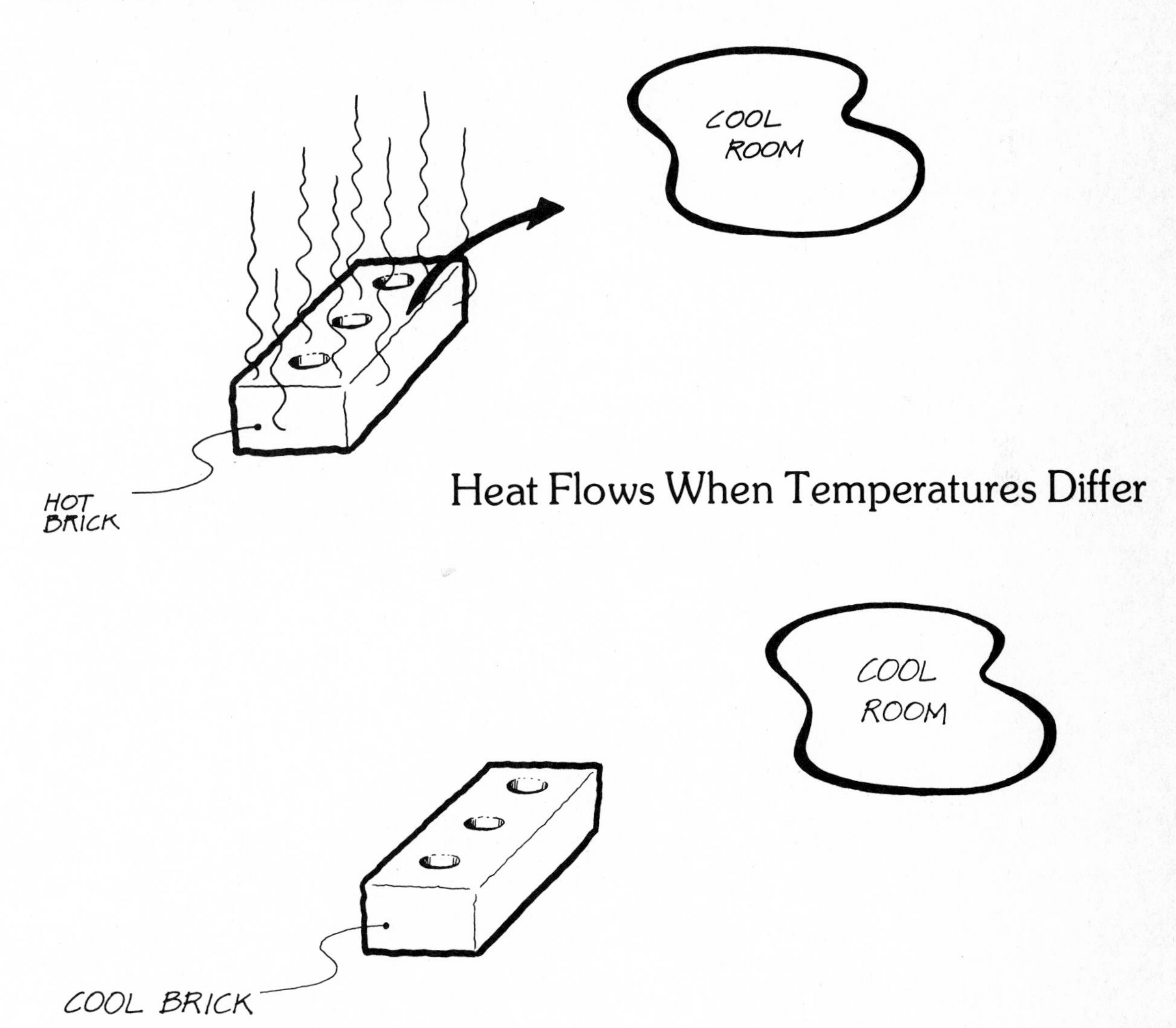

Heat Flows When Temperatures Differ

No Heat Flows When Temperatures Are Equal

Similarly, a full can of water will leak volume from a hole in the side of the can. The depth of the water is higher than the depth of the hole, so the ***depth difference*** drives volume out through the hole. Eventually, all the volume that can leak out does so. When this happens, the water depth has fallen so that it is the same as that of the hole. There is no more depth difference, so no more volume flows out through the hole. Just as a difference in temperature causes heat to flow, so a difference in depth causes volume to flow. When there is no temperature difference, heat flow ceases; when there is no depth difference, volume flow ceases.

Volume Flows When Depths Differ

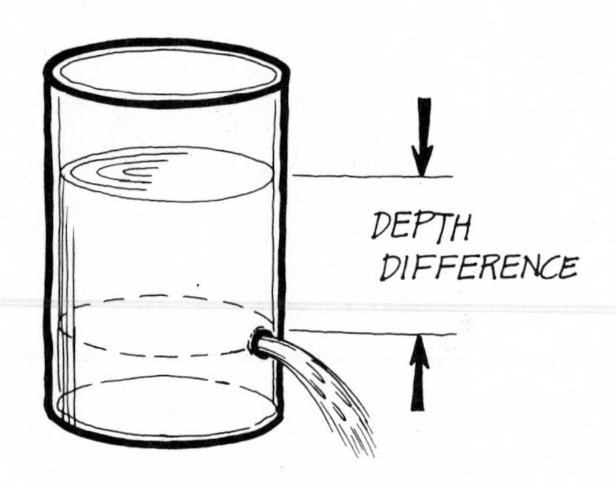

No Volume Flows When Depths Are Equal

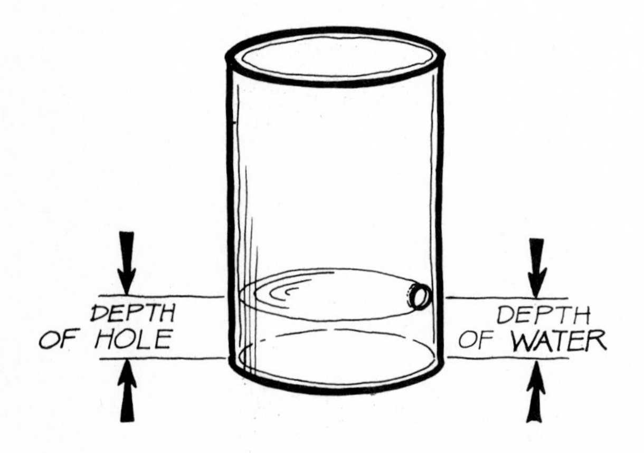

The *direction* of volume flow is always downhill—from a higher depth to a lower one. If two cans having different depths are connected by a tube, volume will always flow *toward* the depth that is lower. Note that the narrow can has much less volume than the wide one. Even so, volume flows toward the can with the lower depth. It's depth, not volume, that causes volume to flow.

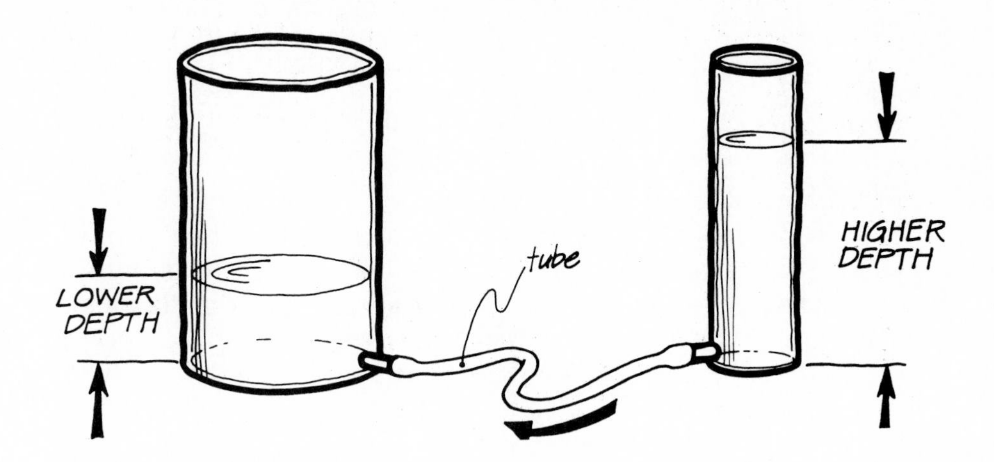

Volume Flows Toward Lower Depth

Similarly, heat flows downhill from a higher temperature to a lower one. If a small, hot stone is put into a big pan of warm water, heat will flow out of the stone and into the water, since the stone is hotter than the water. Heat flow is always *toward* the object with the lower temperature—in this case, the water. Even though the big pan of water has much more heat than the hot little stone, heat still flows from the stone into the water. It's the temperature that makes the heat flow, not the amount of heat each object has.

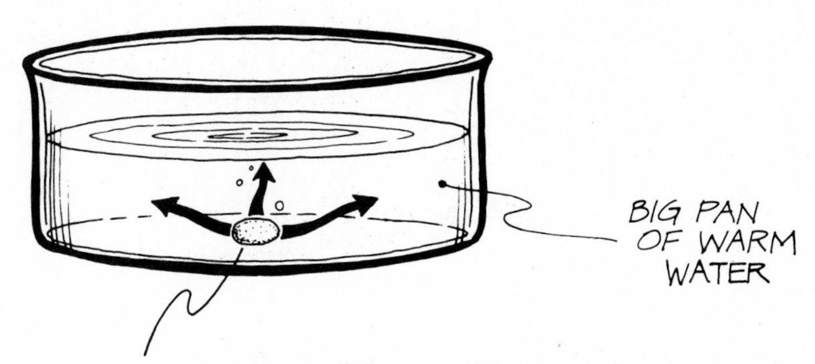

Heat Flows Toward Lower Temperature

Just as volume flows downhill toward lower depths, so heat flows downhill toward lower temperatures. Only a depth difference (not volume) can cause volume to flow, just as only a temperature difference (not heat) can cause heat to flow.

You may think it strange that heat flow is the same at cold temperatures as at hot temperatures. But consider this example. Suppose we took a frozen chicken out of the freezing compartment of a refrigerator and put it into the main part of the refrigerator. The chicken would warm up—heat would flow into it—even though it was in a cold refrigerator. Since the chicken started out colder than the refrigerator, heat had to flow into it to get it to be as "warm" as the refrigerator. If we then put the chicken into a hot oven, more heat would flow into it until it became as hot as the oven. In this example, heat flow depends only on the temperature difference between the chicken and its surroundings, not on whether the surroundings are hot or cold.

Heat Flows Only Because Temperatures Differ

By analogy, the same volume flows from a hole in a can whether the can is on the floor or a table. Only the depth difference between the water level and the hole is important, not whether the can itself is raised or lowered.

Volume Flows Only Because Depths Differ

THERMAL RESISTANCE 3

Temperature differences are what cause heat to flow: heat flows from hot things to cold things. But other factors determine *how fast* the heat will flow. These other factors, when lumped together, cause a *thermal resistance.* The bigger the thermal resistance, the harder it is for heat to flow, since the resistance to the flow of heat is increased. *Resistance* is common to other forms of flow as well: electrical resistance restricts the flow of electric current, and fluid resistance restricts the flow of volume.

Fluid resistance might be thought of in terms of the size of a hole in the side of a container: a little hole has more fluid resistance than a big hole. The little hole doesn't let much volume out, but a big hole will, even though both holes are at the same depth. The little hole "resists" the flow of volume through it more than the big hole, so it has a high resistance.

Fluid Resistance: A Hole in Can

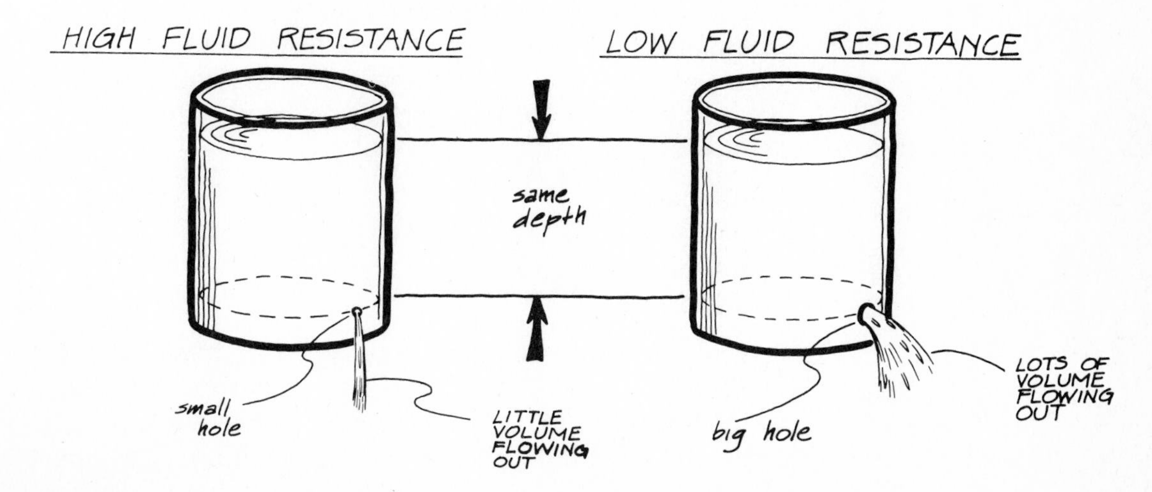

Similarly, *thermal resistance* is a measure of how hard it is for heat to flow. Sometimes we say one house is better insulated than another house, and that's exactly what thermal resistance is—how well insulated something is. Since a well-insulated house has a higher thermal resistance than a poorly insulated one, the well-insulated one loses less heat than the poorly insulated one. The temperature difference may be the same for both houses (room temperature inside and cold outdoors), yet heat will leak at a slower rate from the one with high thermal resistance—the well-insulated one.

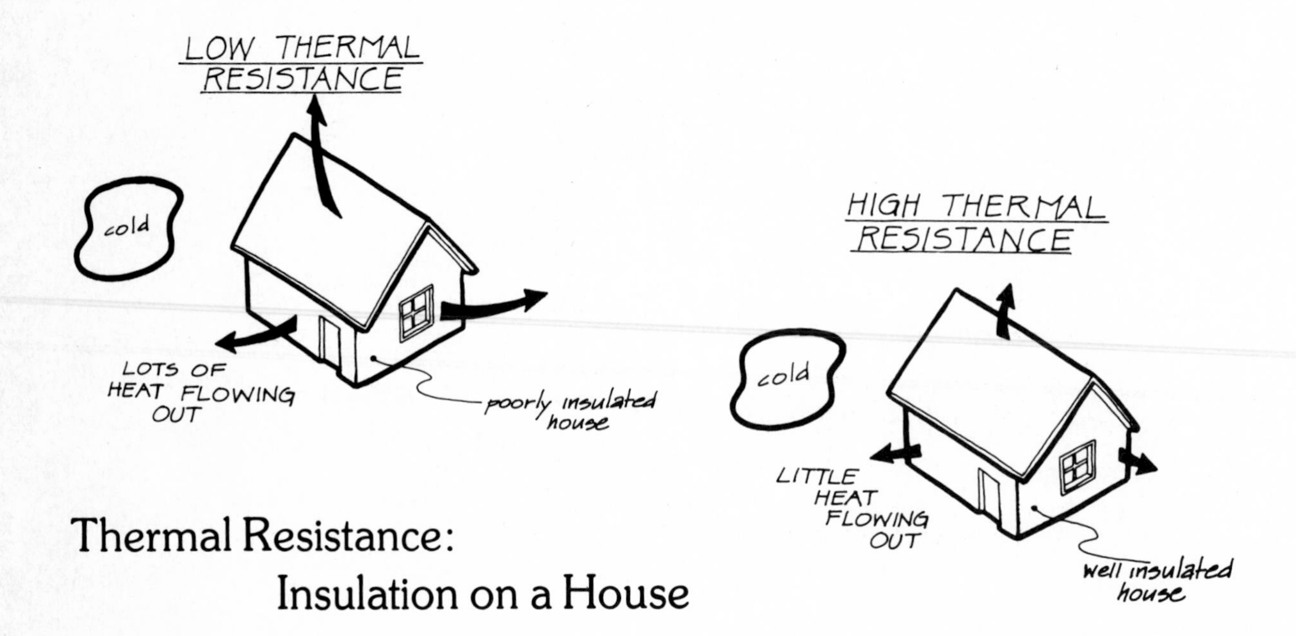

Thermal Resistance: Insulation on a House

The extent of thermal resistance is caused by several factors, depending on the way heat moves from place to place. There are four important types of thermal resistance, corresponding to each of the four important ways in which heat moves in solar heating systems. They are:

1. conduction
2. convection
3. radiation
4. transport

Before we look at each of these types of heat flow in detail, first let's see how they differ generally. As we have learned, heat is like water in that it flows from place to place. If it flows through material that isn't moving, the heat flows by means of *conduction* heat transfer. For instance, a silver spoon is hot when you've been stirring coffee because heat flows easily through the silver from the hot coffee to your fingers. The silver itself doesn't move, but heat flows through it. When we show conduction we'll always use a straight arrow to distinguish it from the other ways in which heat flows.

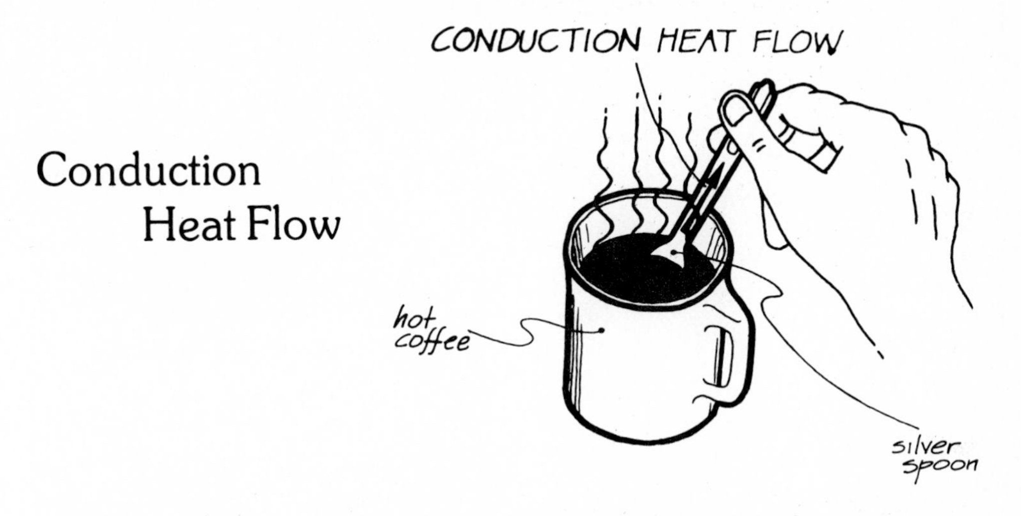

In *convection heat flow*, a surface heats a liquid or gas near it, and the heat is carried away by the liquid or gas. For example, you feel cold on a windy day because the wind carries heat away from your skin by convection heat transfer. We'll use a curved arrow ( ) to show convection heat flow.

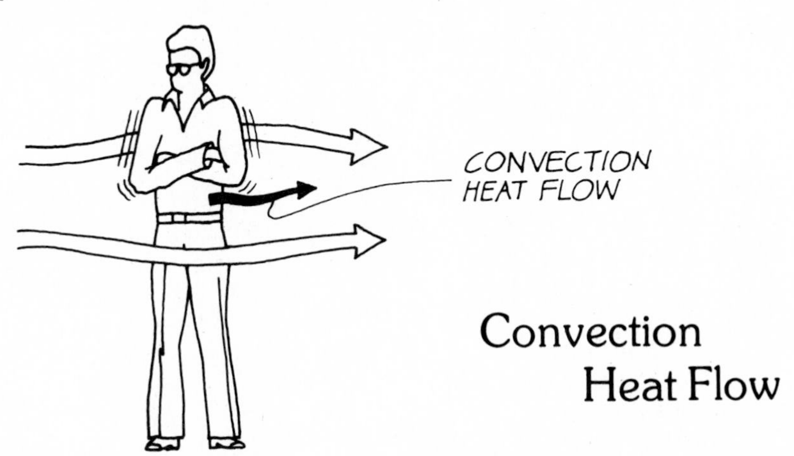

Radiation heat flow is a special kind of energy that travels like radio waves through air—and even through a vacuum. You feel warm in front of a fireplace mostly because the flames and hot coals move heat to your skin by radiation. Wavy arrows ( ) will be used to show heat flowing by means of radiation.

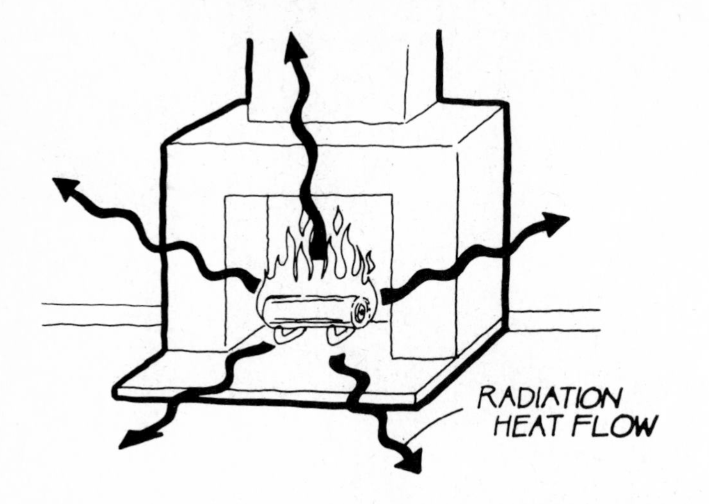

Radiation
Heat Flow

Transport heat flow is similar to convection in that it involves the transport of a heated liquid or gas. But instead of a surface heating the liquid or gas, previously heated liquid or gas is transported, or moved, from one place to another. Hot gases rising out of a chimney is a form of transport heat flow. We'll use a curlicue arrow ( ) to indicate transport heat flow.

Transport
Heat Flow

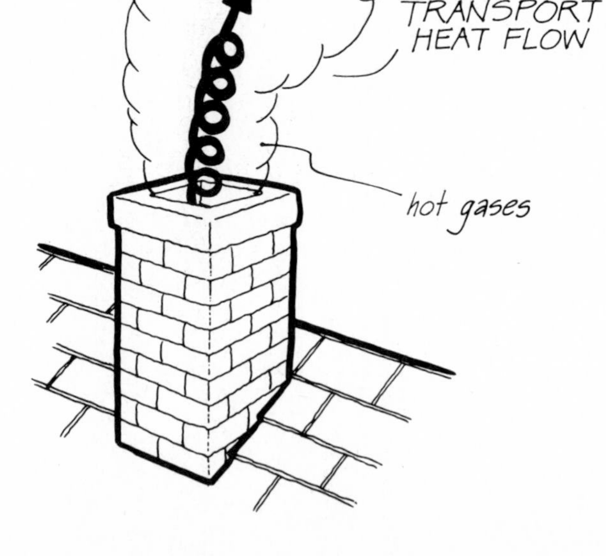

Now that you have an idea of the ways heat flows in general, let's take a more detailed look at each way.

CONDUCTION AND CONVECTION 4

First, let's look at *conduction.* Two kinds of materials are involved in conduction heat flow: *conductors* and *insulators*. Conductors let heat flow through them easily, but insulators prevent heat from flowing. Most metals, such as silver, copper, aluminum, and steel, are good conductors of heat. Insulators, on the other hand, are usually lightweight materials, such as straw, fiberglass batting, or plastic foam. Other materials, such as glass, concrete, rubber, wood, and dirt, are neither good conductors nor good insulators; heat passes through them more easily than through an insulator but not as easily as through a conductor.

Conduction heat flow depends on several factors. We've already learned that it depends in the first place on temperature difference (as does all heat flow) as well as on the kind of material involved (whether conductor or insulator). It also depends on the *area* of the *heat-flow path.* In the example of the silver spoon, the heat-flow path is from the coffee to the spoon's end.

Conduction
Heat Flow Path

To understand how the heat-flow path changes the conduction heat flow through an object, suppose we try to melt an ice cube using a cup of hot coffee. If we put one end of a silver bar into a coffee cup and hold an ice cube against the other end, the ice will hold one end of the bar at a fixed temperature (since ice melts at 32° F); how quickly the ice melts will measure how much heat is flowing through the bar from the coffee. A thick bar will melt the ice more quickly than a thin one because a thick bar lets more heat from the coffee through—the heat-flow path of the thicker bar has a bigger area.

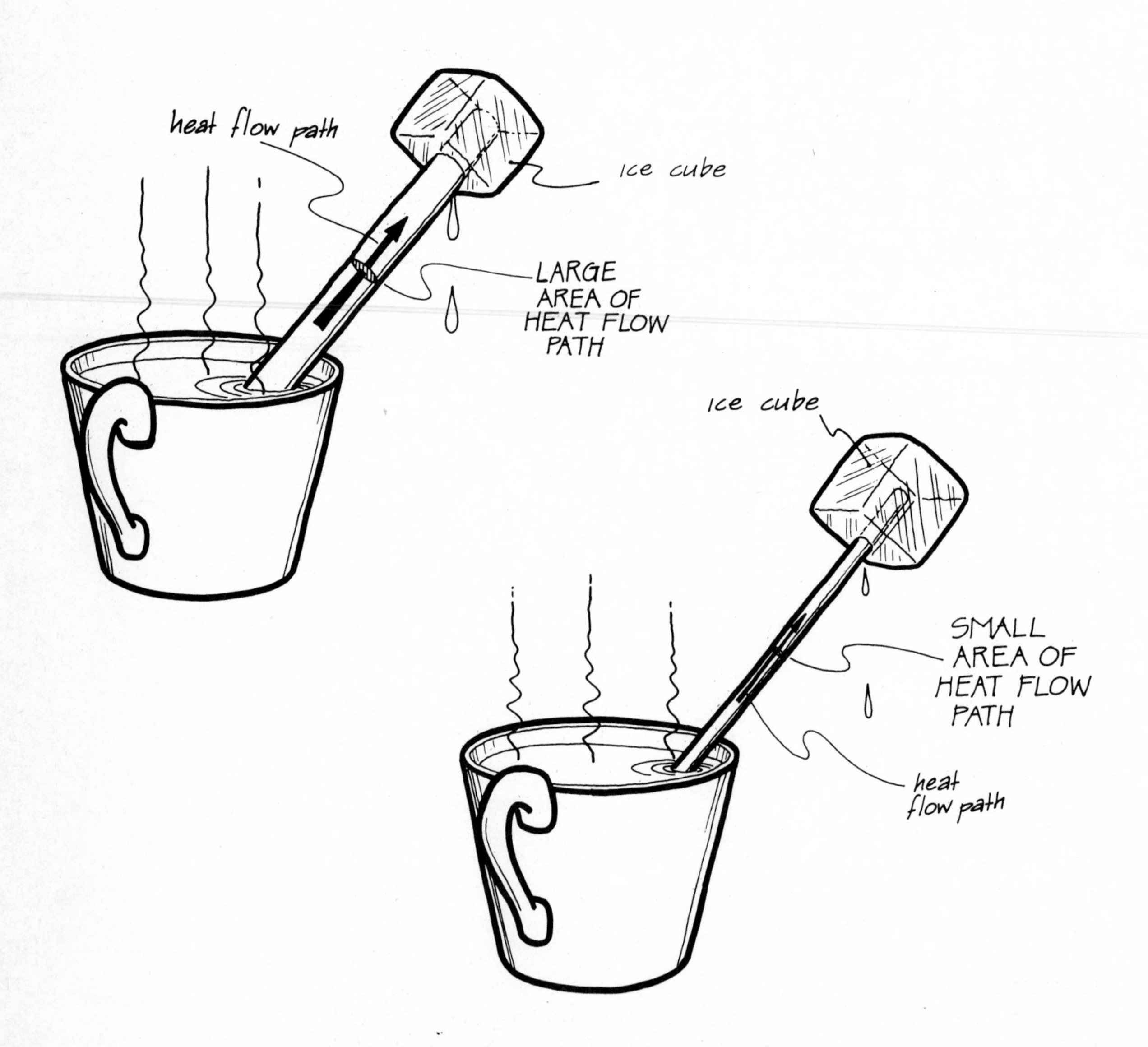

Thick Bar Conducts More Heat

The length of the heat-flow path is also important. Less heat flows through a longer bar than through a shorter bar because the length of the heat-flow path impedes the flow of heat.

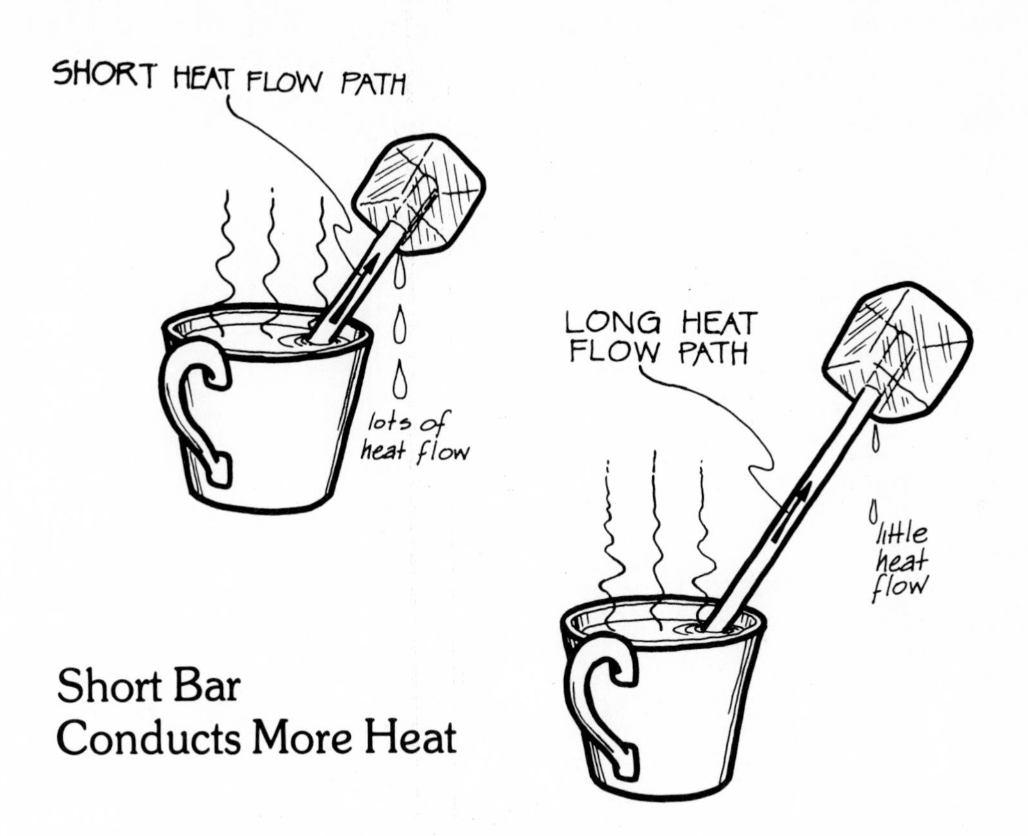

Short Bar Conducts More Heat

Four factors affect how much heat will flow through a material by conduction: temperature difference, material, area, and the length of the heat-flow path. By analogy to water flow, think of a pipe attached to the bottom of a big tank.

Four factors also affect how much fluid will flow through the pipe. As shown in the following figure, lots of volume will flow out of a tank of water through a short, fat pipe. But depth difference affects the volume flow: if the tank isn't full, less will flow out. Second, the material also has an effect. Less volume flows if the tank is filled with honey than if it's filled with water. Third, flow-path area affects volume flow; there is less flow through a thin pipe than a fat one. Fourth, the length of the flow path is important, since less water flows through a long pipe than through a short one.

Volume Flow and Conduction Heat Flow

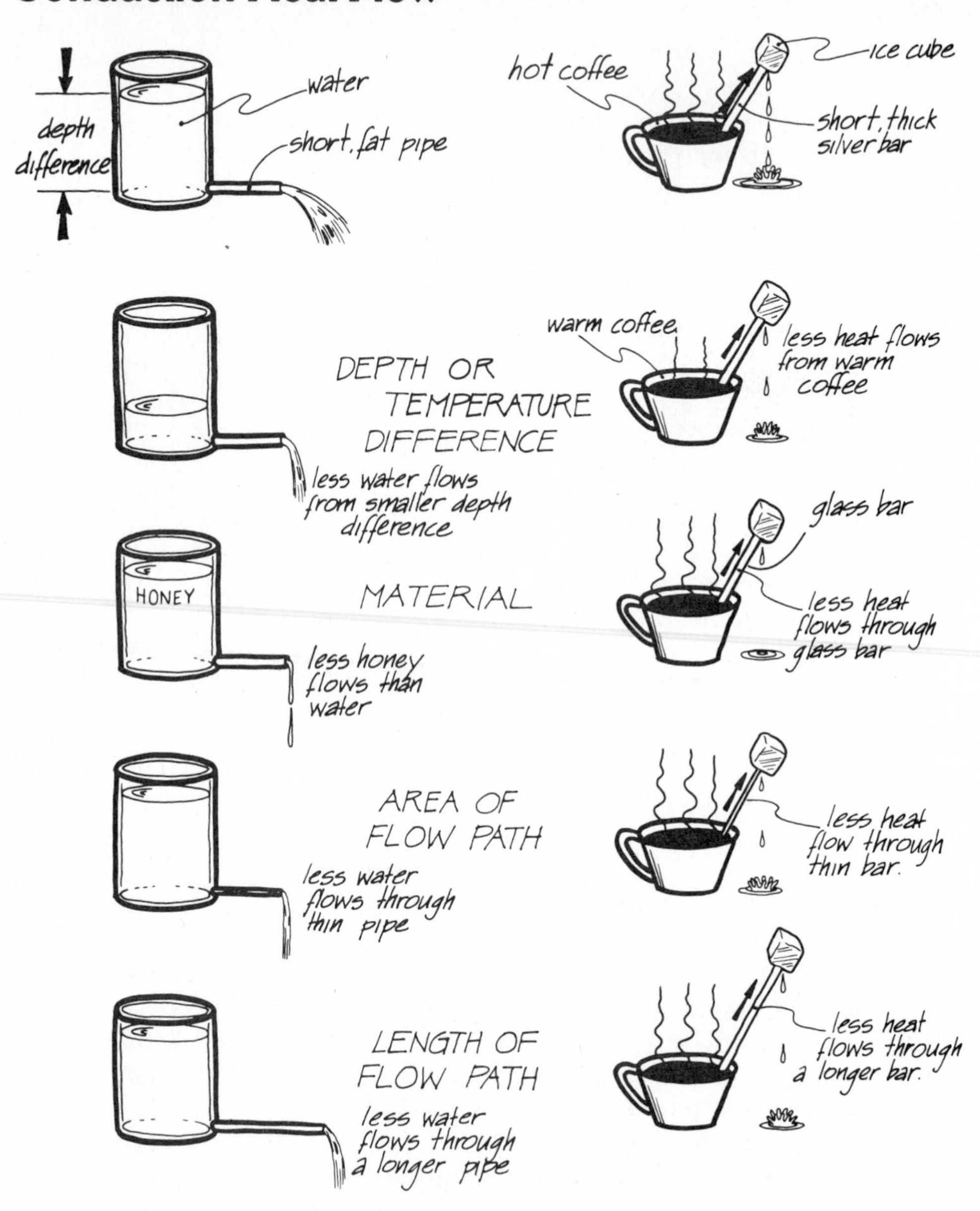

Now let's discuss *convection*. Heat flow by convection occurs when a gas such as air or a liquid such as water flows by a surface. In the example used earlier, the surface was your skin as air blew by it. As with conduction heat flow, convection heat flow depends on several factors. One factor, we've already learned, is temperature difference. In convection heat flow, it's the temperature difference between the surface and the convecting gas or liquid that's important. Other important factors are:

1. the area of the surface
2. the speed of the air or water over the surface
3. whether it's gas or liquid that flows by the surface.

More Convection from Large Surfaces

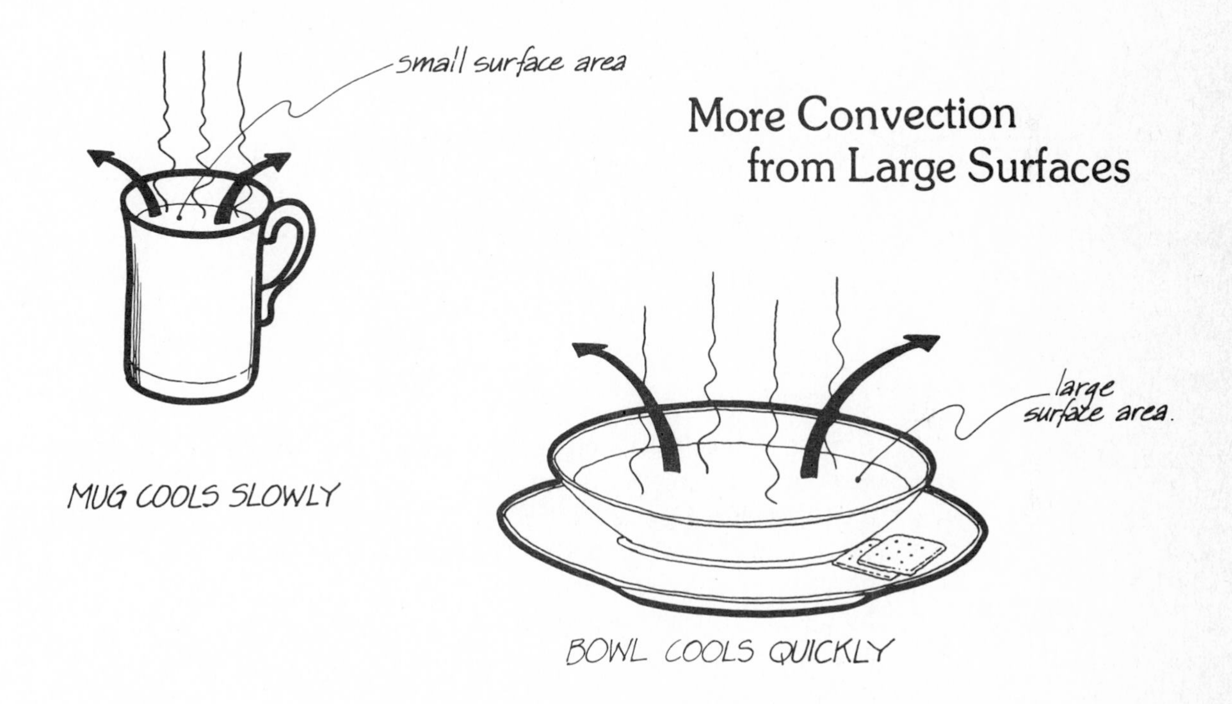

The larger the surface area is, the more convection heat is lost from the surface. If you have the same amount of soup in a bowl as in a mug, the bowl will cool faster than the mug because its surface area is larger than the mug's. Second, the speed at which the convecting gas or liquid flows by the surface influences how quickly the heat flows. Suppose two bricks are placed in an oven, heated to the same temperature, and then taken out. One is placed on a table in still air; the other is placed in front of a fan. The one sitting in still air would cool much more slowly than the one sitting in front of a fan. You might think that the brick in still air wouldn't have any air flowing by it at all. Actually, air is flowing by—but much more slowly than if the fan were blowing air by it. The brick heats the air near it, which gets lighter and rises and is replaced by cooler air, which in turn is warmed and rises. This phenomenon is called *natural convection*, since no fan is needed to cause the air to move: the air rises past the surface simply because the surface is hot.

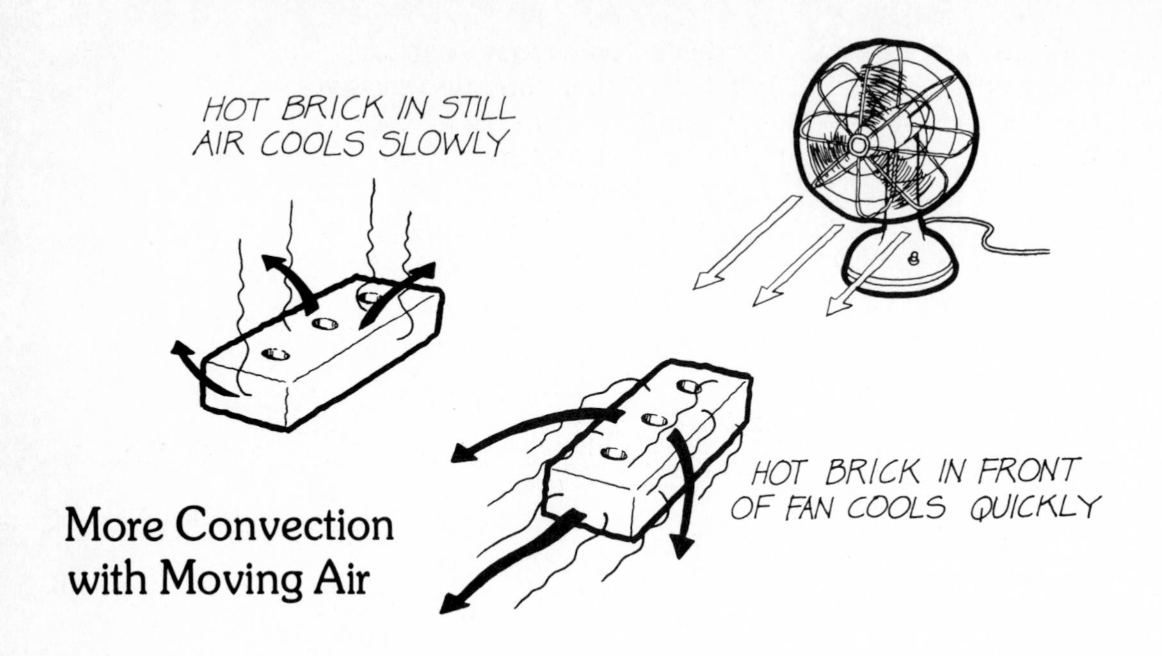

More Convection with Moving Air

Smoke rising from a cigarette illustrates how natural convection removes heat from burning tobacco; the smoke moves with the air, showing how a hot object causes air to move by it. Conversely, cold surfaces also cause convection, but in the opposite direction. Convection currents *fall* from a cold object, as you may have noticed when opening the freezer door to the refrigerator.

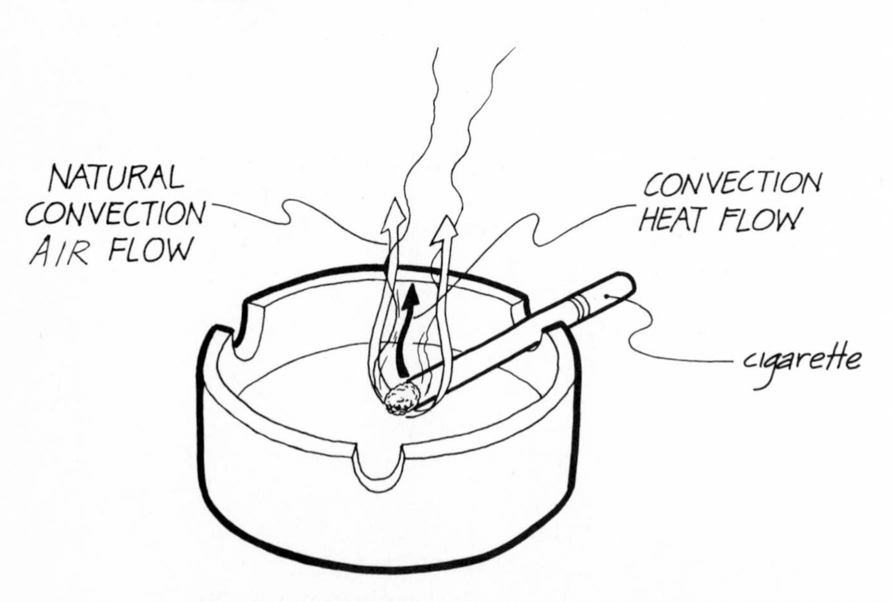

Natural Convection in a Cigarette

The last factor that is important in convection heat flow is whether a gas or a liquid is involved. Liquids are much better than gases at causing heat to flow from a surface. Eggs cook in a few minutes in boiling water, but they would take many times longer to cook in an oven at the same temperature. The eggs gain heat slowly by air—a gas—convecting over them, but they heat up quickly when water—a liquid—is convecting over them.

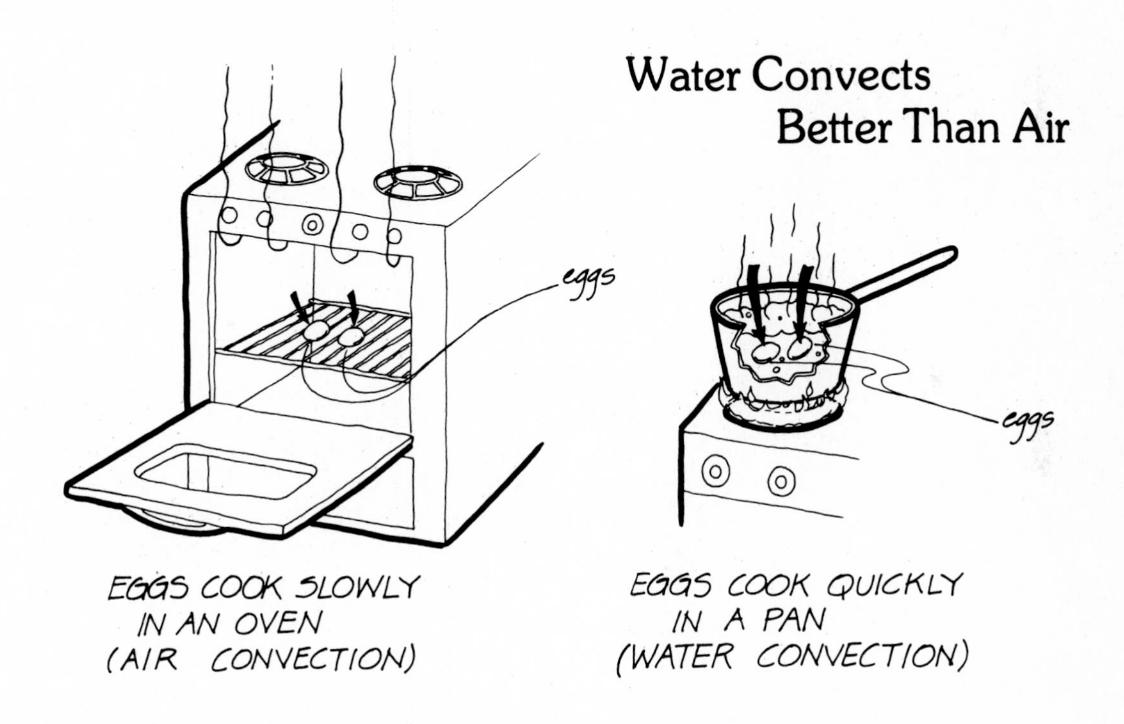

In summary, convection heat flows fastest when there are big temperature differences of large surfaces with liquid flowing by them quickly. Little convection heat flows with small temperature differences of small surfaces in still air.

RADIATION AND TRANSPORT 5

Radiation heat flow is very similar to convection heat flow in many ways. Radiation heat flows from surfaces, as does convection heat flow. Radiation is also similar to convection in another basic way: the bigger the surface is, the more heat flows from the surface. Most surfaces lose heat by both radiation and convection, and it's sometimes hard to separate the two kinds of heat flow. For instance, the example used to illustrate convection—a cooling bowl and a cooling mug of soup—applies equally to radiation. The bowl loses heat quickly by both convection and radiation because the bowl has a bigger surface area. Some of the heat flow from both the bowl and the mug takes place by means of convection, and some by radiation.

However, radiation differs significantly from convection, especially with regard to solar energy. One big difference is that convection depends on a liquid or gas flowing by the surface. With radiation heat flow, the heat leaves the surface even if there is no gas or liquid around. In the vacuum of space, where no gas or liquid exists, heat flows from a space satellite only by radiation. Radiation heat flow is often called *infrared* radiation. Infrared film, for example, is sensitive to radiation heat flow; it is used to photograph heat leaving objects by radiation. We'll learn more about radiation heat flow when we discuss solar radiation—a kind of radiation heat flow that comes from the sun.

A second important difference between radiation and convection is that radiation heat flows from a surface in straight lines. If you put your hand to the side of a heating coil on an electric range, you would feel the warmth of the radiation heat flow. But if you put a book or even a piece of paper between the coil and your hand, it would block the radiation heat flow. Likewise, if you're sitting in front of a fire, the radiation heat flow from the embers is blocked if someone sits in front of you.

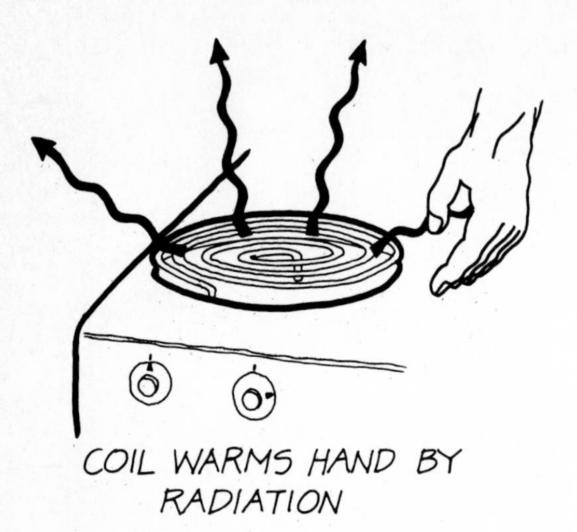

Radiation Heat Flow is Easily Blocked

As with all heat flow, radiation heat flow depends on temperature difference. The important temperature difference is that which the surface sees. In the case of the bowl, the soup's surface sees the temperature of its surroundings (the room), so the important temperature difference is that between the soup and the room. In the case of the hand, the skin's surface "sees" the coil, so the important temperature difference is that between the skin and the coil.

Shiny surfaces play a role in radiation heat flow because they reflect radiation. For example, a thermos bottle is "silvered" to reflect back any heat that the hot soup inside might

Shiny Surface Reflects Radiation Heat Flow

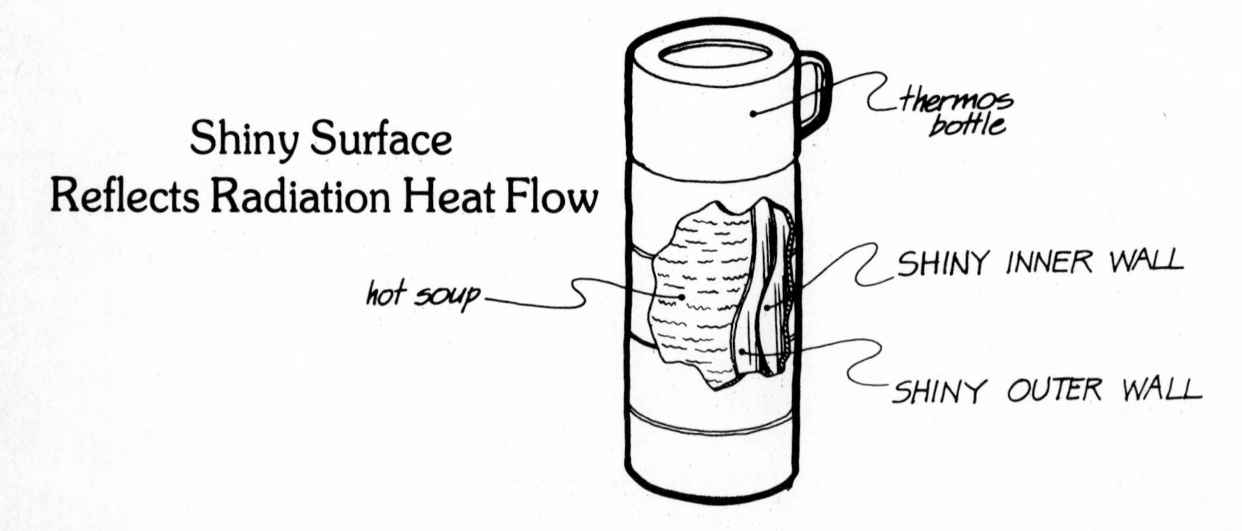

lose by radiation. This trick keeps the soup hot longer. Fiberglass batting used to insulate houses also has a shiny reflecting surface to keep heat in.

Radiation heat flow, then, is much like convection in one way: heat flows from surfaces at a rate depending on the size of the surface. But unlike convection, radiation heat flow travels in straight lines and can be very easily blocked from flowing. In addition, radiation heat flow can be affected by how shiny the surface is, whereas convection cannot.

Finally, let's discuss *transport* heat flow. Heat flow via a flowing gas or liquid is somewhat different from conduction, convection, and radiation. Suppose you wanted to heat up a cup of coffee by adding more hot coffee to it. When you add the coffee you're also adding heat, since the hot coffee has heat stored in it. Heat is physically transported from the pot to the cup by transferring a certain amount of hot coffee from the pot to the cup.

Heat Flowing by Transport

In most transport heat flow, the flowing gas or liquid moves continuously, not just for a short time, as in the example of the coffee and cup. To understand this, think again of the example of a house losing heat through its chimney. In this case, the flowing medium which transports the heat is a gas—mostly hot air. But for as much hot air that goes up the chimney, there must be as much cold air that comes in from outside to replace it. The cold air isn't so obvious because it leaks in through cracks in the walls, windows, and doors. Nevertheless, as much cold air leaks in as hot air leaves. The air, then, moves continuously: it enters in through the cracks, is heated by the

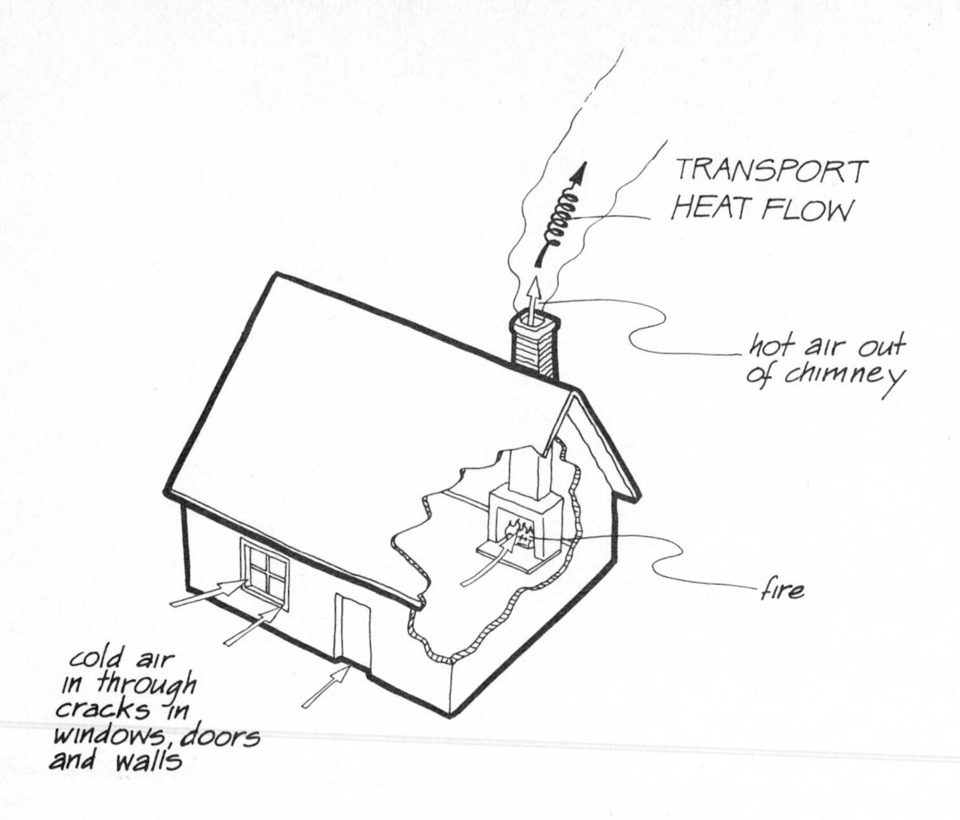

Transport Heat Flow up a Chimney

fire, and flows out through the chimney. As in conduction, convection, and radiation, the amount of heat flow depends on the temperature difference. In this case, the important temperature difference is that between the incoming airflow and the outgoing airflow—the temperature of the hot air going out the chimney minus that of the cold air leaking in through the cracks in the house.

Transport heat flow depends on two other factors besides temperature difference. One is how fast the gas or liquid is flowing, and the other is what the flowing medium is. To better understand the first factor, suppose your house were heated by hot-water radiators. Hot water flows in one end and out the other end; how fast it flows depends on how wide you open the valve. To get lots of heat out of the radiator, you open the valve so that lots of hot water will flow through the radiator. If you don't want much heat, you close the valve so that little hot water will flow into the radiator; it, in turn, will add little heat to the room.

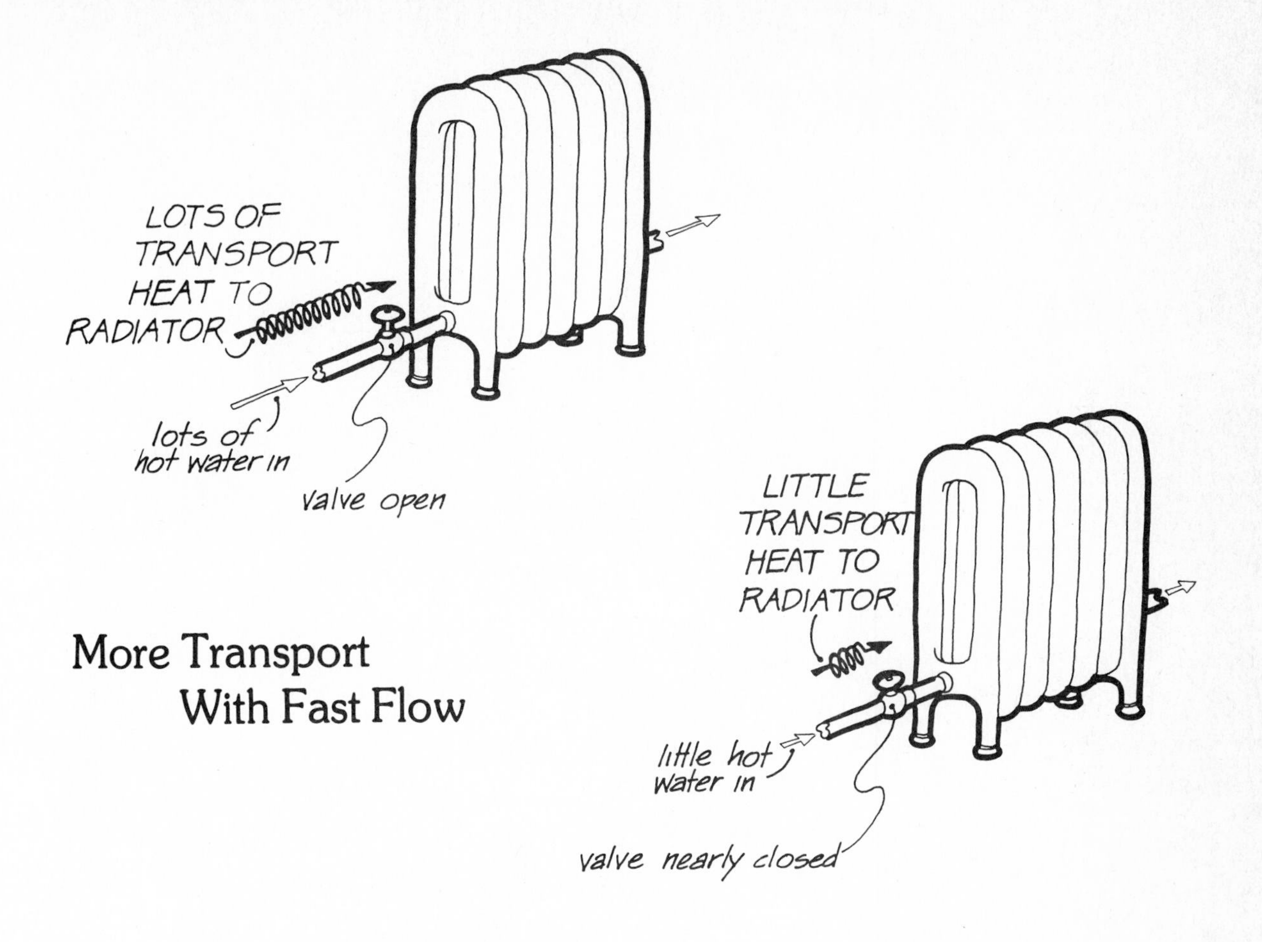

More Transport With Fast Flow

The second factor in transport heat flow is the flowing medium itself. Air and water are the two most important *heat transfer mediums* used in solar heating systems today. The key difference is the volume needed. In this respect, water is a much better heat transfer medium than air. More than a thousand times more air than water is needed to transport the same amount of heat. Fans are often used to transport heated air, while heated water is generally moved by pumps.

THERMAL STORAGE 6

In solar energy heating systems, it's not enough to simply capture the sun's heat—it must also be stored. While heating a house during the day is useful, the house needs heat at night as well. Heat collected during the day must be stored until night. How is heat stored?

Storing heat is analogous to storing volume. If a cylindrical tank is filled with water, the amount that it holds—the volume stored—depends on how high its depth is. Similarly, the heat stored in a substance depends on how high its temperature is; a hot brick stores more heat than a warm one.

The Deeper, the More Volume Stored; The Hotter, the More Heat Stored

If a solar-heated house is to be kept at 70° F during the winter, heat stored at 50° F isn't useful. Only if the temperature of the storage material is above 70° F can the stored heat be used to warm the house. Thus, useful heat in terms of room heating is heat that is stored in a substance that is hotter than the room's temperature. By analogy, the useful water in a tank is the water stored above the outlet. Once the depth drops below the outlet, the remaining volume won't flow out, so it's not useful.

TANK ONLY STORES USEFUL VOLUME ABOVE OUTLET TAP

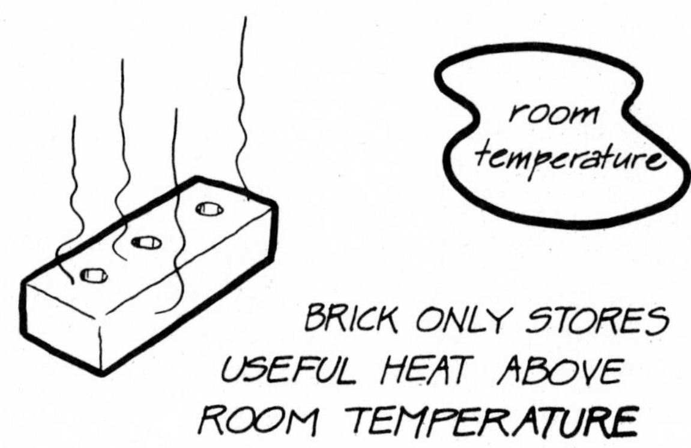

BRICK ONLY STORES USEFUL HEAT ABOVE ROOM TEMPERATURE

Not All Volume and Heat is Useful

The amount of heat stored also depends on how much material is heated. The more storage material there is, the more heat can be stored at the same temperature above room temperature. Similarly, a bigger tank holds more volume than a smaller tank when both are of the same depth.

MORE VOLUME STORED

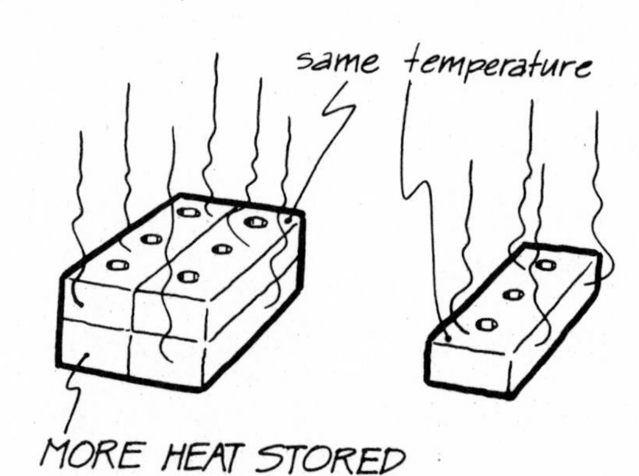

MORE HEAT STORED

Storage Depends on the Amount of Storage Material

Often the *amount of material* is measured by volume. Using volume as a basis of measurement, different materials can be compared as to how well they store heat. In solar heating systems, heat is usually stored in water or crushed rock. Water stores heat somewhat better than does rock. A container of water can store two to three times the heat as the same-sized container filled with crushed rock.

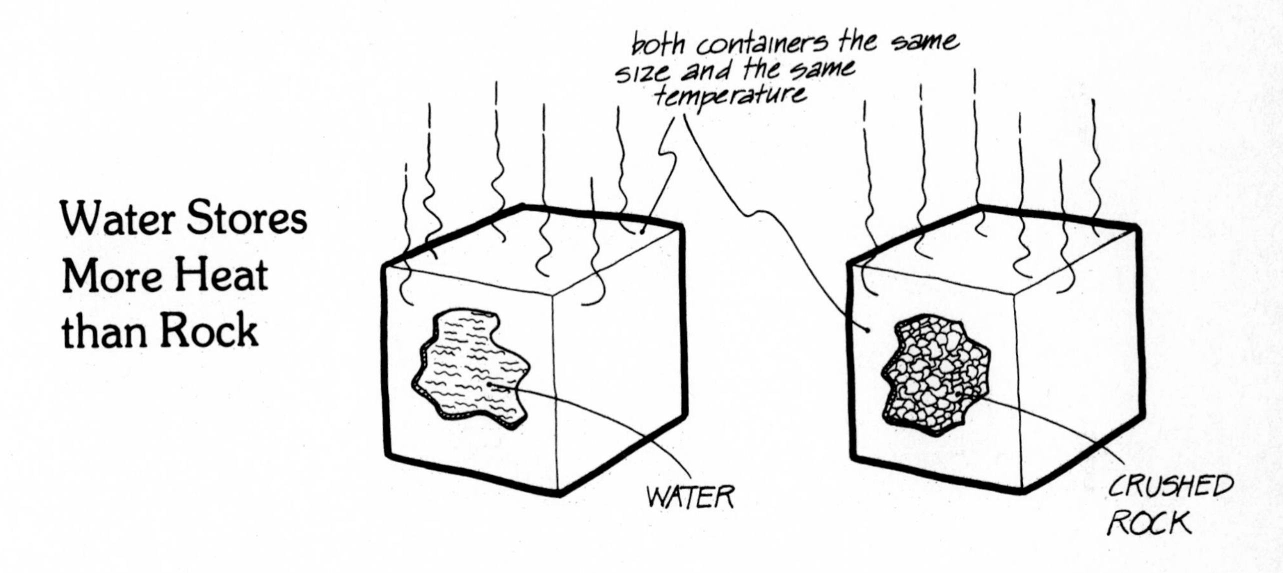

Some materials store heat very poorly. Air, for example, stores almost no heat. A container of air stores one-thousandth the amount of heat as the same-sized container of water does.

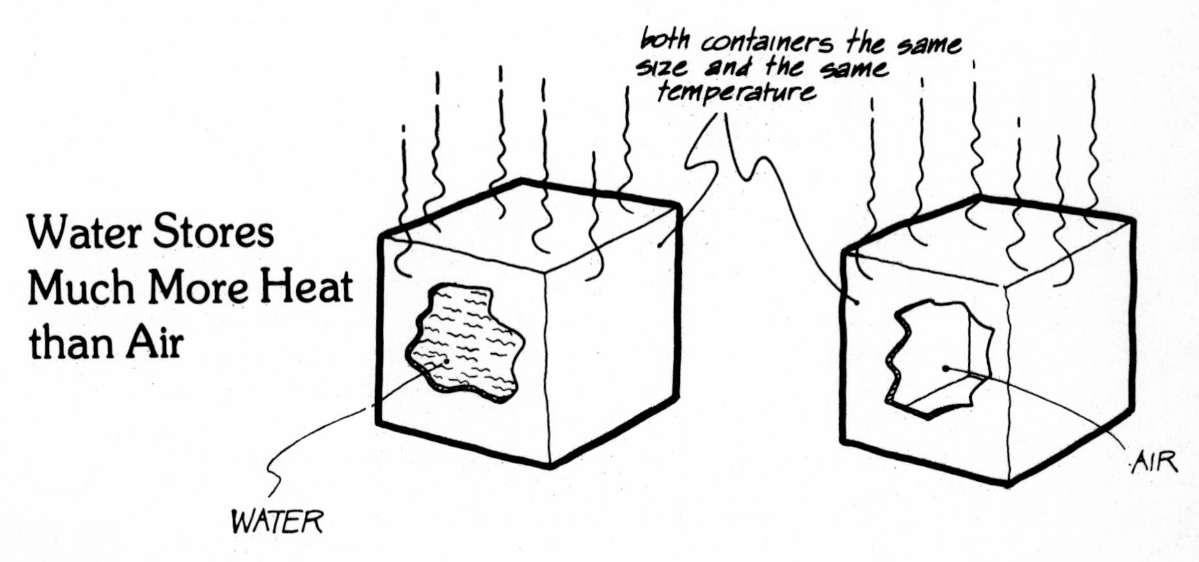

To understand heat storage, consider the following example. When you turn off your furnace on a cold day, your house doesn't get cold immediately because heat is stored in the walls, the floor, and the furniture (very little is stored in the air). The stored heat leaks out slowly and keeps you warm for a short time even when the furnace is off. Adobe houses in the Southwest have very thick walls that store the day's heat well into the night.

An Adobe House Stores Heat

Another form of heat storage is called *latent* heat storage. Instead of storing heat simply by getting hotter, some materials store heat by melting. Ordinary wax is such a material. Let's take a piece of solid wax. As we heat it, it gets hotter as does any other material. But when it gets so hot that it begins to melt a curious thing happens: even though we continue to add heat to the wax, it doesn't get any hotter. All the heat goes into melting the wax, and it stays at nearly the same temperature until all the wax is melted. Melted or liquid wax again acts like any other material—the more heat added, the hotter it gets.

When melted wax loses heat, its temperature naturally drops. But when it begins to solidify again it stops getting cooler. As it solidifies, it gives back all the extra heat—the latent heat—that was put into it when it was melted. While the wax is solidifying, its temperature remains nearly constant. Once it has completely hardened, it once again behaves like any other material, dropping in temperature as it gives up heat.

Suppose we use wax to store heat. As the wax melts, heat will be stored. Later, if we want the stored heat, we'd let the wax solidify again, giving up its latent heat. Materials like wax are called *phase change material* because they store heat by changing phase—that is, by melting or solidifying.

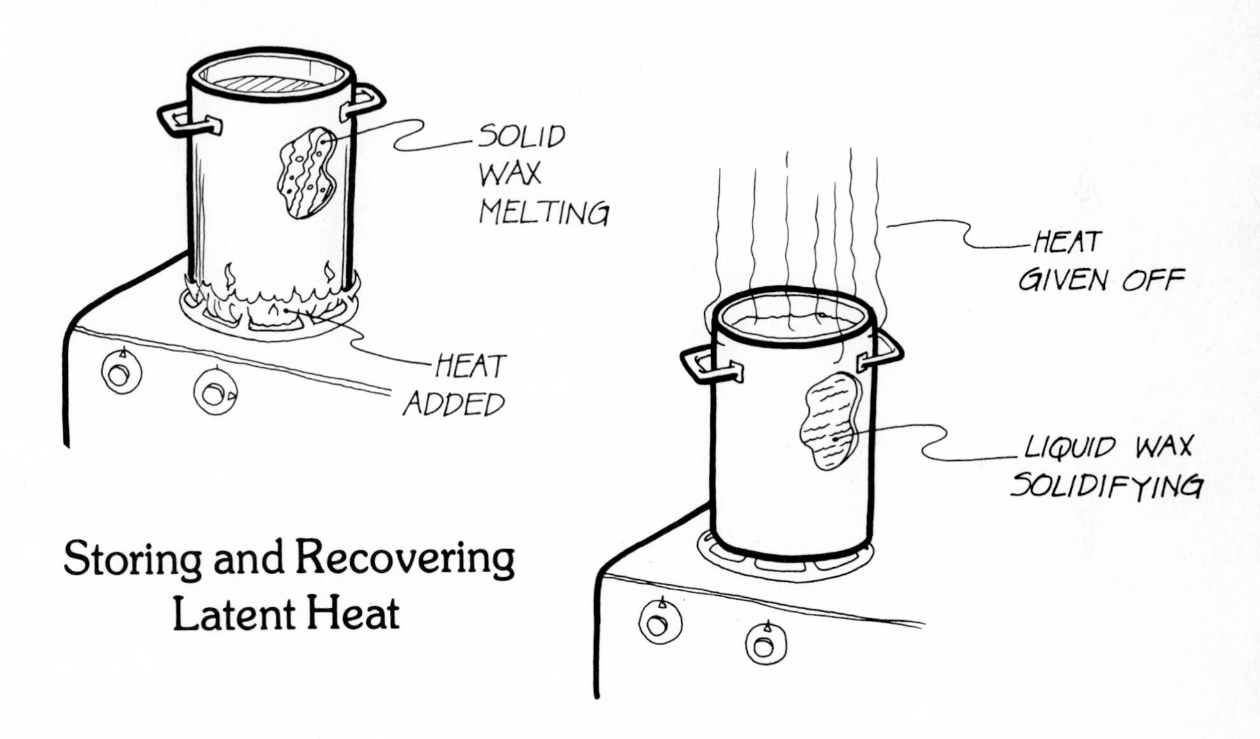

Storing and Recovering Latent Heat

Why would you go to the trouble to use wax to store heat rather than water or crushed rocks? Because wax and other phase change materials can store a lot of heat. Wax can store several times the amount of heat of an equal volume of water and many times the heat of an equal volume of crushed rock. Wax itself is fairly expensive to use to store heat. Other phase change materials are cheaper and don't pose a fire hazard that wax might.

The water analogy to a phase change material is a tank that has a bulge in its sides. When the water depth is below the bulge, adding volume only increases the depth, as it would in any other tank. But when the depth reaches the bulge, a lot more volume is needed to raise the depth. Extra volume is stored in the bulge just as extra heat is stored in the wax when it melts. If the tank is drained the extra volume can be recovered, just as the latent heat in the wax can be recovered when it solidifies. When the water depth in a tank is at the level of the bulge, volume can be added or taken away without changing the depth much. Similarly, when wax is near its melting temperature, heat can be added or taken away without changing the wax's temperature much.

Tank with Bulge is Like Latent Heat Storage

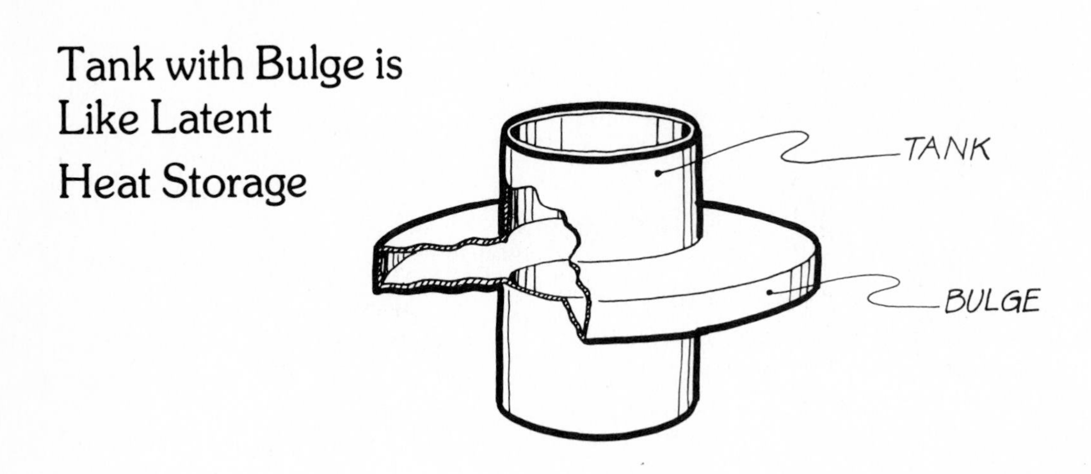

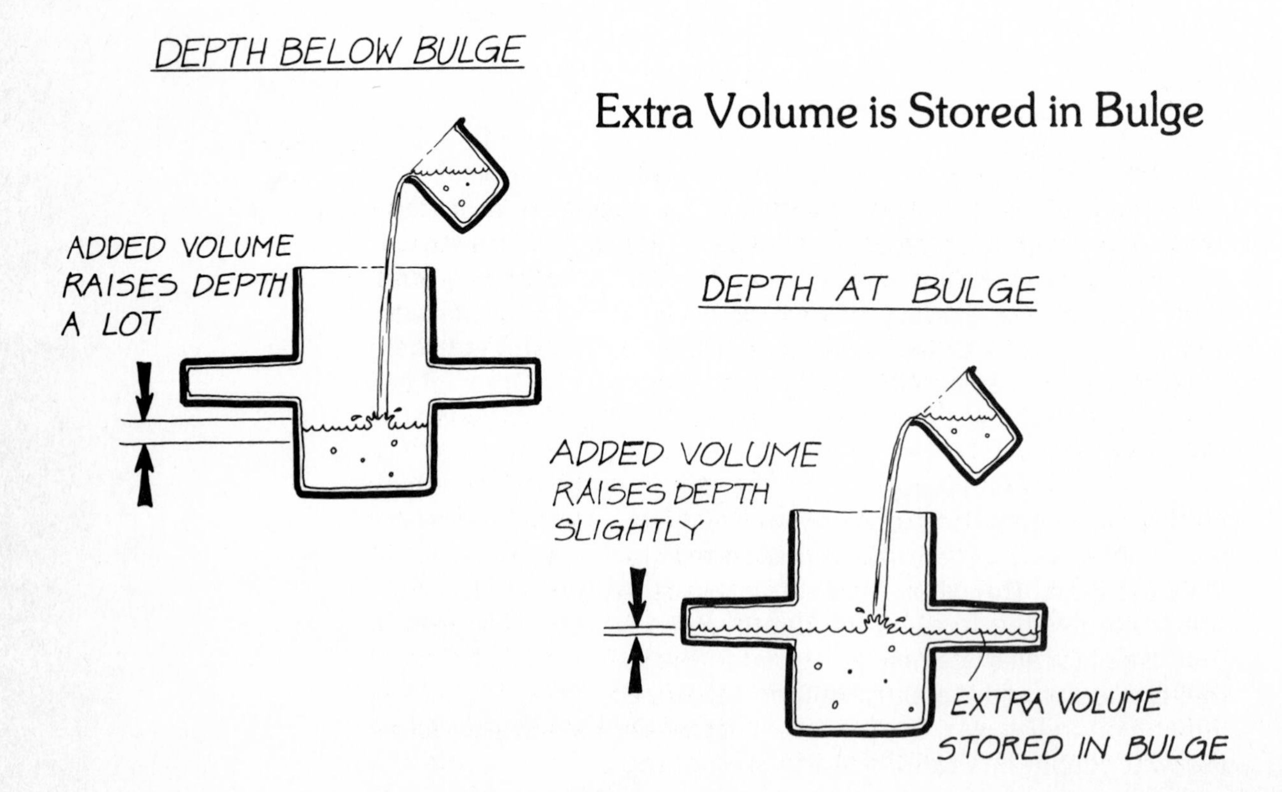

Extra Volume is Stored in Bulge

HOUSE HEATING 7

We know that heat can be transferred by several different methods and that heat can also be stored. Let's see how *all* the different types of heat flow and storage combine together in a familiar application. Later, when we talk about solar heating systems, one important use will be in heating a house: the solar heating system will add heat and the house will lose heat. The heating losses of a house provide a good example of the way heat can flow by several methods at the same time.

Before we can talk about *house heat losses*, we have to understand *house heating*. A house's heating system—whether it is gas, oil, electric, or even solar—tries to keep the house at a more or less fixed temperature. When the house gets too cold the furnace comes on and adds heat to the inside of the house. The heat might be added by means of baseboard electric heaters, a gas-burning hot-air system, an oil-fired water furnace delivering heat through baseboard radiators, or any number of other heating systems. The added heat, by whatever system, tends to heat the rooms—to increase their temperature. Once the indoors temperature gets a little hotter, the thermostat sends a signal to the furnace and it turns off.

Once the furnace turns off, the house slowly loses heat to the outdoors. It doesn't lose heat right away, because heat is stored for a time in the walls and floors. When the thermostat senses that the house has become too cold, it turns the furnace on again and the cycle repeats. You usually don't notice these changes in temperature in the house because they are so small. For example, if you set your thermostat at 70° F, the furnace might turn on when it cooled to 68° F and turn off again when it reached 72° F. Room temperature is always moving slightly up and down—getting hotter and colder—but it always remains very close to the set temperature.

Indoors Temperature Controls Furnace

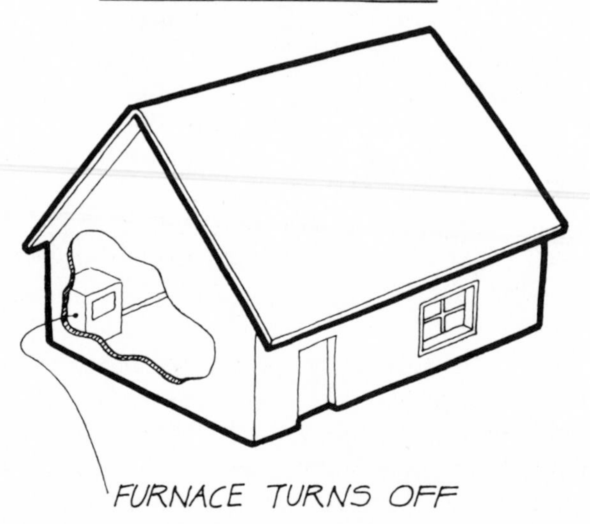

Now, remember the water analogy to heat and temperature. The temperature of an object is equivalent to the water depth of a tank; the heat is equivalent to the volume of water; and the heat flow is equivalent to the water flow. How could we make the *water equivalent* of a house-heating system? Suppose we let water pour from a faucet into a bucket that has small holes in its bottom. The faucet is equivalent to a house's furnace because the faucet adds water to the bucket just as a furnace adds heat to a house. The holes in the bucket represent the way the heat leaks from the house. When the faucet adds water to the bucket, it leaks out through the holes; when the furnace adds heat to the house, it leaks out through the walls, doors, windows, roof, and cellar.

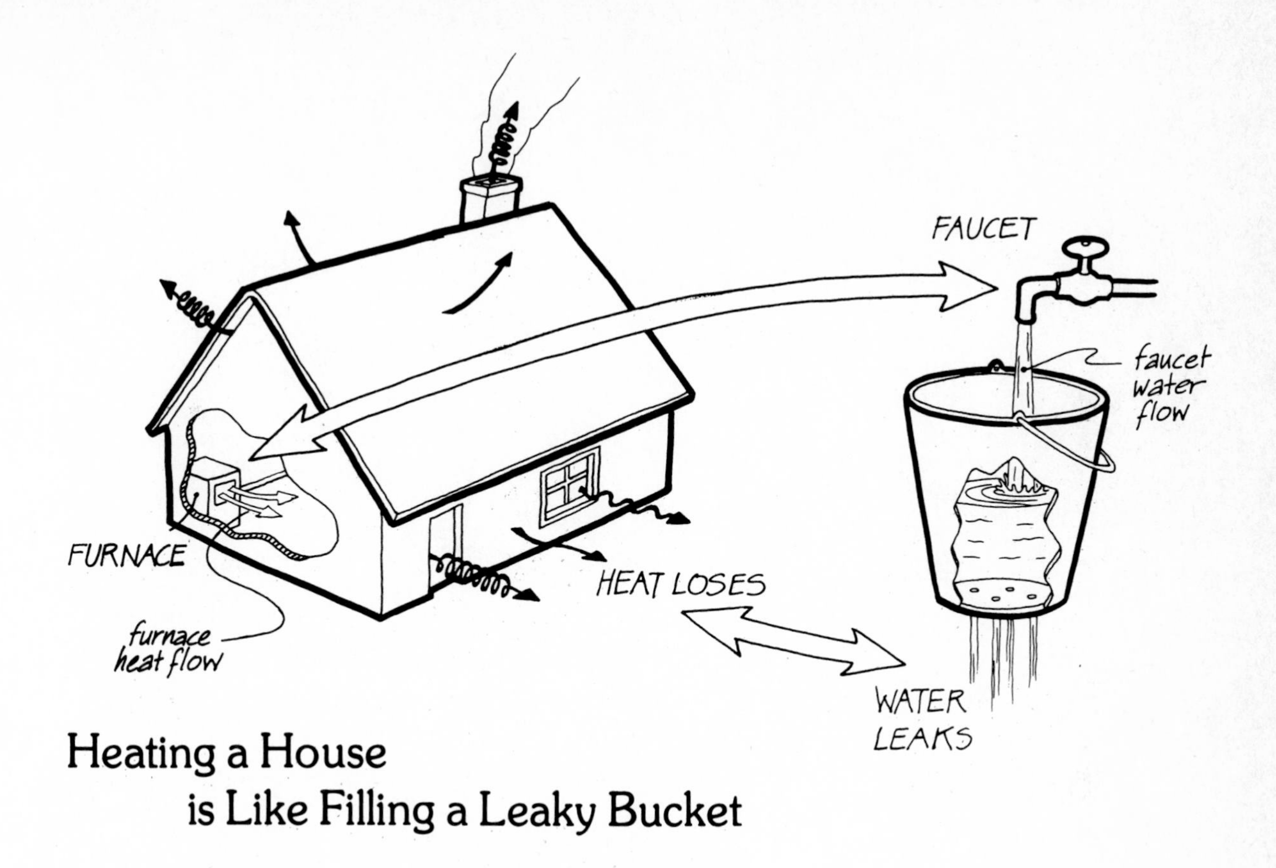

Heating a House is Like Filling a Leaky Bucket

The depth of water in the bucket is equivalent to the room temperature in the house. While the temperature difference is what drives the heat from the house, the *depth difference* is what drives the water from the bucket. The depth difference is simply the depth of water in the bucket, since the holes are at the bucket's bottom. The bottom of the bucket is equivalent to the outdoors temperature.

Just as the thermostat keeps the house's temperature relatively stable, you could keep the depth of the water in the bucket relatively constant. Suppose you were to draw a line on the bucket indicating the depth you wanted to hold. If the depth dropped *below* the line, you'd turn on the faucet. Similarly, when a thermostat senses that the house is getting a little too cold, it turns on the furnace. The faucet adds water to the bucket just as the furnace adds heat to the house. When the depth went *higher* than the line on the bucket, you'd turn off the faucet. Likewise, when the thermostat senses that the house's temperature is getting a little too hot, it turns off the furnace. After a while, the water would leak out of the holes in the bucket and the depth would drop below the line. It wouldn't happen very

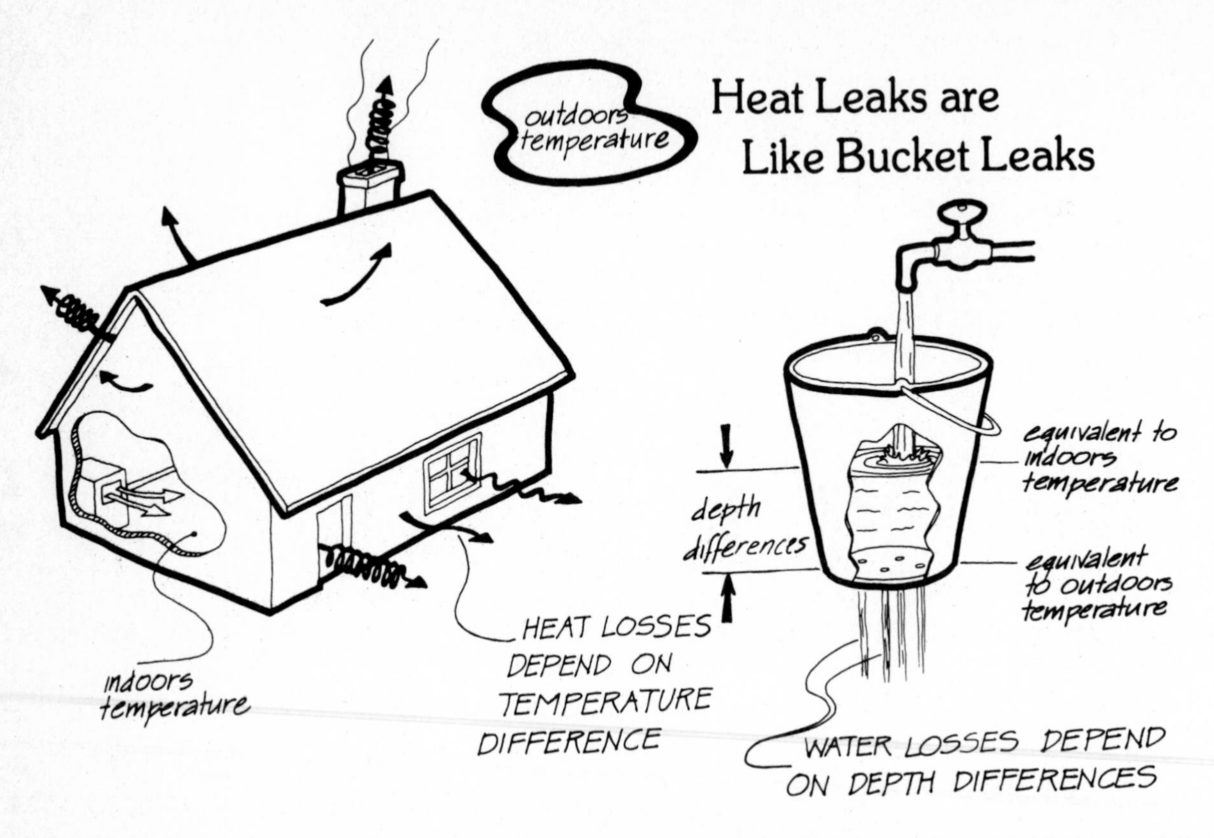

fast, because the bucket stores volume and it would take a while for the volume to leak away. Similarly, the heat flows from a house through all of its heat leaks, and the house slowly drops in temperature as it loses heat. When the water goes below the marked line, you'd again turn on the faucet and begin the cycle again.

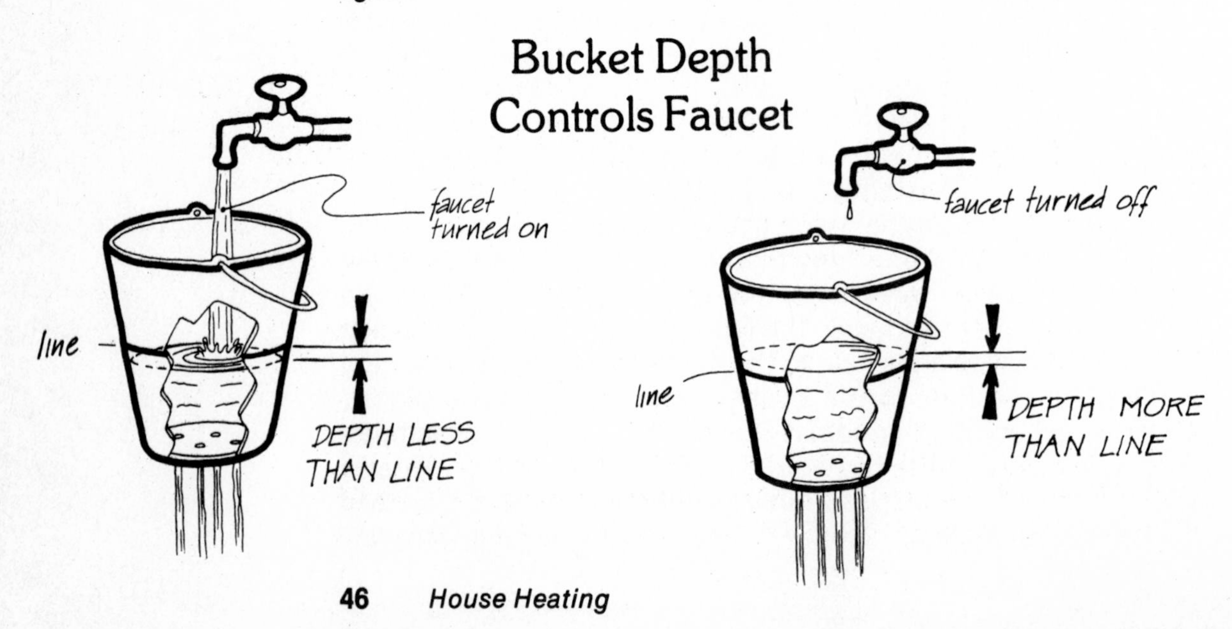

If you were very diligent, you'd be able to keep the depth pretty close to the line on the bucket by turning the faucet on and off properly. Since you were holding the depth more or less constant, the water lost through the holes would be more or less constant. If you held the depth constant for a long time, say an hour, the total volume that poured into the bucket would equal the volume that leaked out during the same hour. Just as the faucet adds as much volume each hour as is necessary to balance the water leakage during that hour, so a furnace adds the same amount of heat that leaks out of the house each hour.

Suppose we set the house's thermostat at 65° F instead of 70° F. What would happen? Heat losses would be less because the temperature difference would also be less. Your heating bill would be less too, because the furnace adds just what the house loses. Similarly, if you marked another line on the bucket at a lower level and kept the depth at this lower line, the water leakage would be less. Since depth differences drive out the water through the holes, less water leaks out. On the average, the faucet adds just as much water as leaks out, so you would use less water at the lower line that at the higher line.

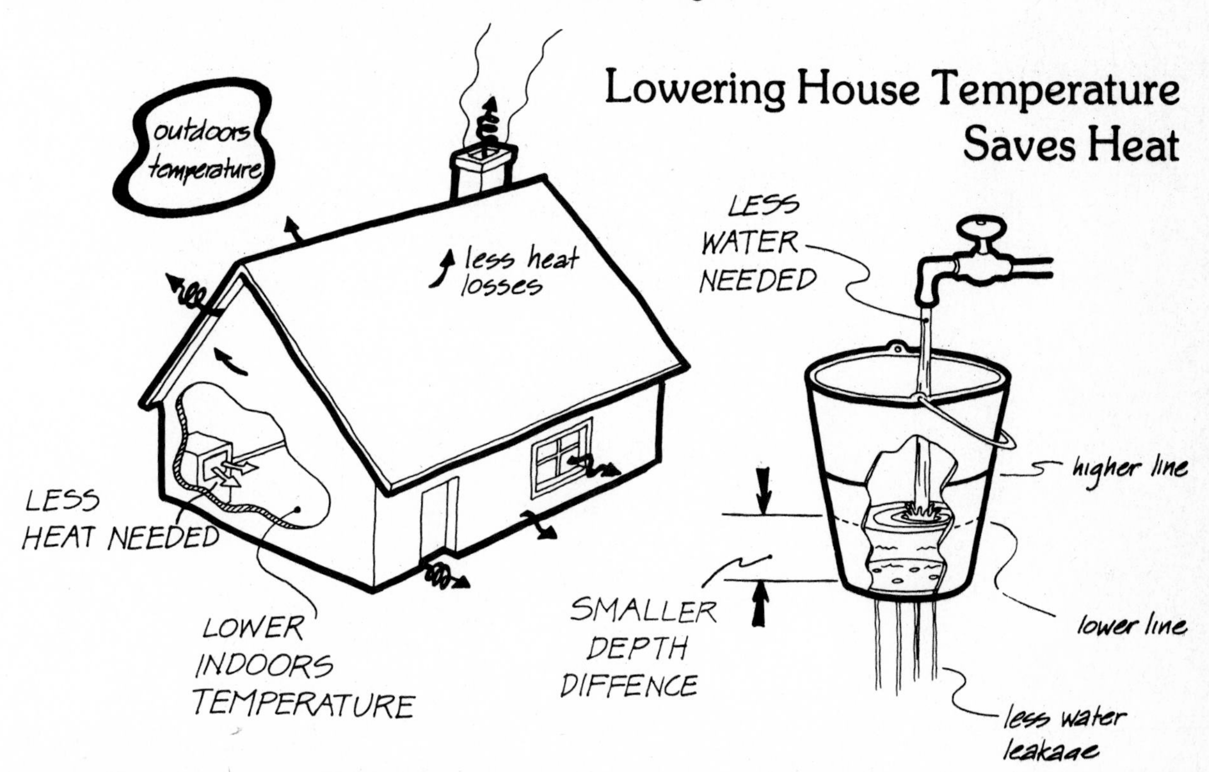

Our country's leaders have urged us to turn down our thermostats to save scarce energy resources—now you have an understanding of why it helps.

HOUSE HEAT LOSSES 8

Houses lose their heat because it's hotter inside than outside. How exactly is the heat lost? Mostly it is lost in three ways: heat leaking by conduction through the walls and roof; heat leaking by radiation and convection through windows; and heat leaking by transport through cracks in windows and doors.

Conduction losses through walls and roofs usually account for about half of the total heat loss in an uninsulated house. But if fiberglass batting or plastic foam is used to insulate the walls and thick fiberglass batting insulates the roof, the conduction losses can easily be cut in half. The thicker the insulation is, the less heat is lost by conduction. A roof with three inches of fiberglass insulation will lose twice as much heat as a roof with six inches of insulation.

Uninsulated House Loses More Heat

The next most important heat loss is through windows. Even though ordinarily very little surface area of a house is made up of windows, heat losses through windows can amount to as much as the heat loss through an insulated wall. If a well-insulated wall of your house is one-tenth glass, as much heat can be lost by conduction through the wall as through windows—even though the wall has a ten times larger area.

Large Heat Loss Through Windows

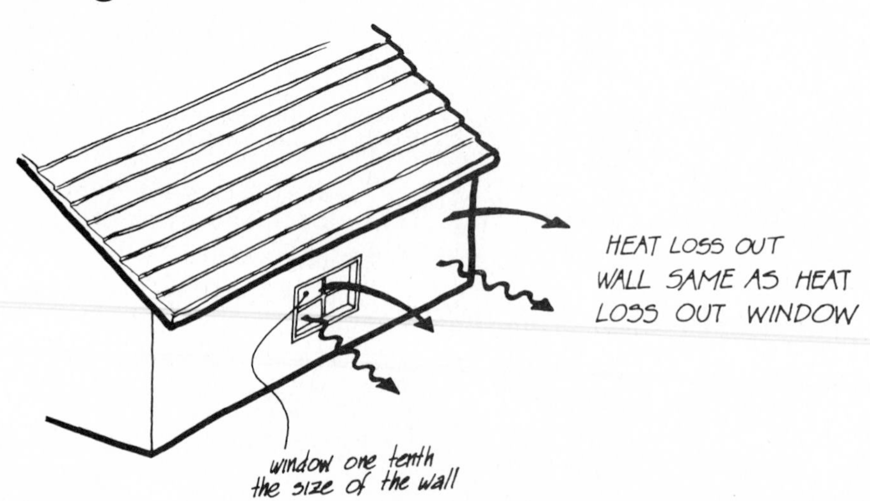

Window heat loss takes place by conduction, convection, and radiation. Heat is convected and radiated to the inside surface of the glass by the warm room. Even though we can see through glass, in terms of radiation heat flow it's as if the glass weren't transparent at all: the glass prevents radiation heat from flowing directly outside. Once the heat reaches the inner surface of the glass, it is conducted through the glass. Glass is a good conductor, so the heat flows easily through the glass to the glass's outer surface. Once there, the heat flows outdoors by means of convection and radiation. Since the wind is often blowing outdoors, heat leaves the window's outer surface very easily.

The net effect of all these combined heat losses is that windows aren't very good insulators. Heat losses through windows can be cut down by using two layers of glass—as in *doubled-glazed* windows or storm windows. A double-glazed window has an air gap between the glass layers, so it's much harder for heat to flow through it. As with glass made of a single layer, it flows by convection and radiation to the window's inner

Single-Glazed Window Insulates Poorly

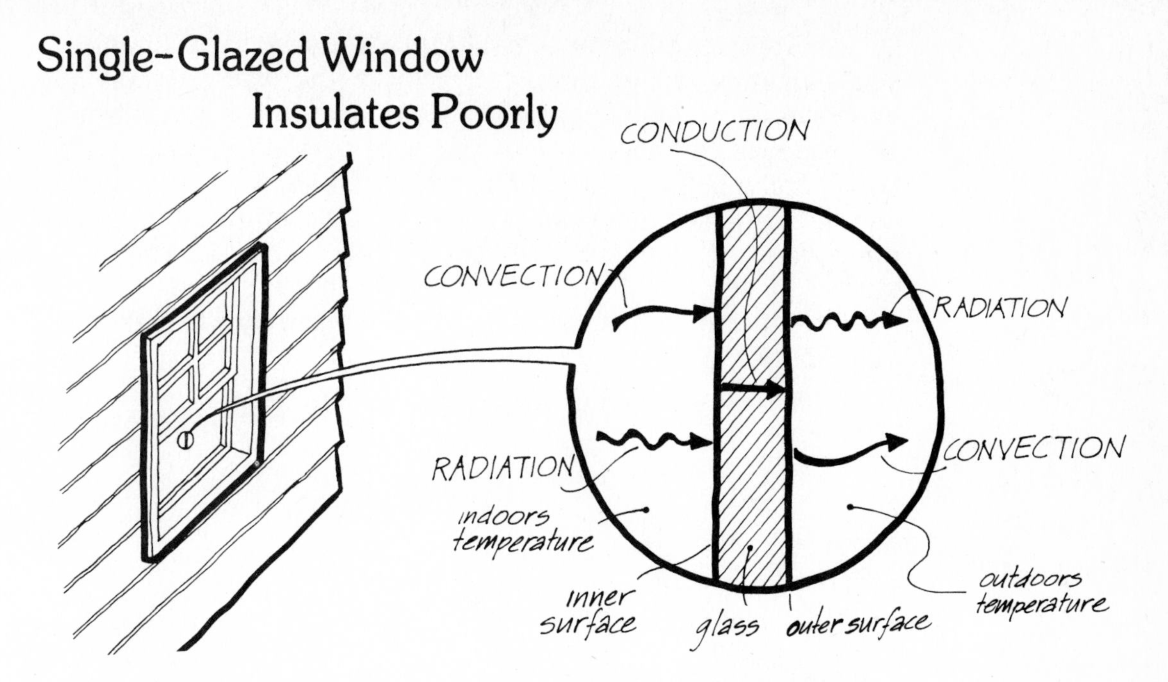

surface. Then it flows by conduction through the first glass layer, by conduction and radiation through the air gap, by conduction through the outer glass layer, and finally by convection and radiation to the outdoors. Double-glazed windows lose about half as much heat as a single-layered window; most of this effect is due to the fact that the air gap acts as a good insulator.

Double-Glazed Window Insulates Better

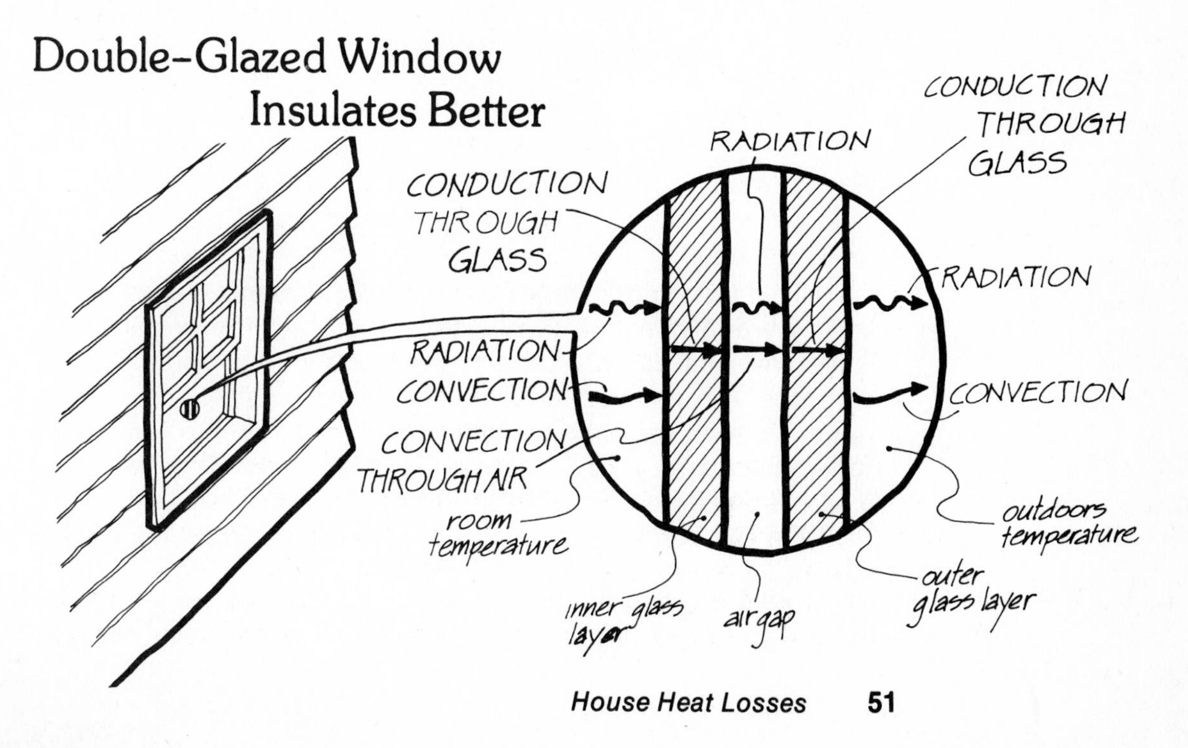

The third major heat loss from a house is by *infiltration*—warm air leaking out through cracks in the walls, doors, and windows and being replaced by cold outdoors air. Infiltration can account for one-third of the heat loss of an uninsulated house and a half or more of an insulated one. Many houses are poorly built, and thus have quite a bit of air infiltration. On the other hand, if a house were *completely* sealed it would feel stuffy. In any case, cutting down on infiltration is an easy matter. Infiltration heat losses can be reduced by weather-stripping doors and caulking windows.

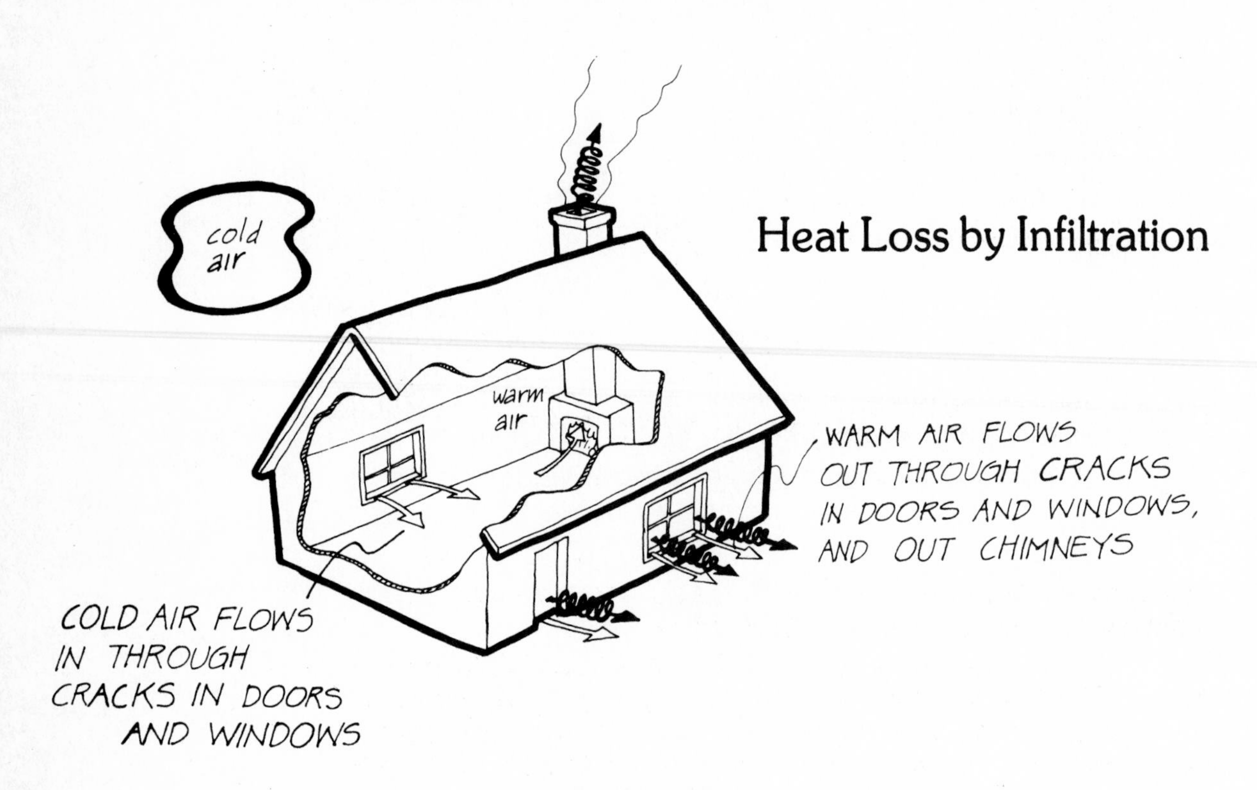

To return to our water analogy. Recall that a house's heat losses are equivalent to the holes in the bottom of a bucket. If we insulate the house—reduce the conduction, window, and infiltration losses—it's like making the holes in the bucket smaller. Just as the furnace needs less heat to keep a well-insulated house warm, so a faucet needs to add less water to a bucket with small holes to maintain a constant water level.

On the average, the furnace adds exactly what heat leaks out of the house. If less heat leaks out, the furnace uses less and your heating bill is less. Similarly, on the average the faucet adds the same amount of water to the bucket as leaks out of it. If

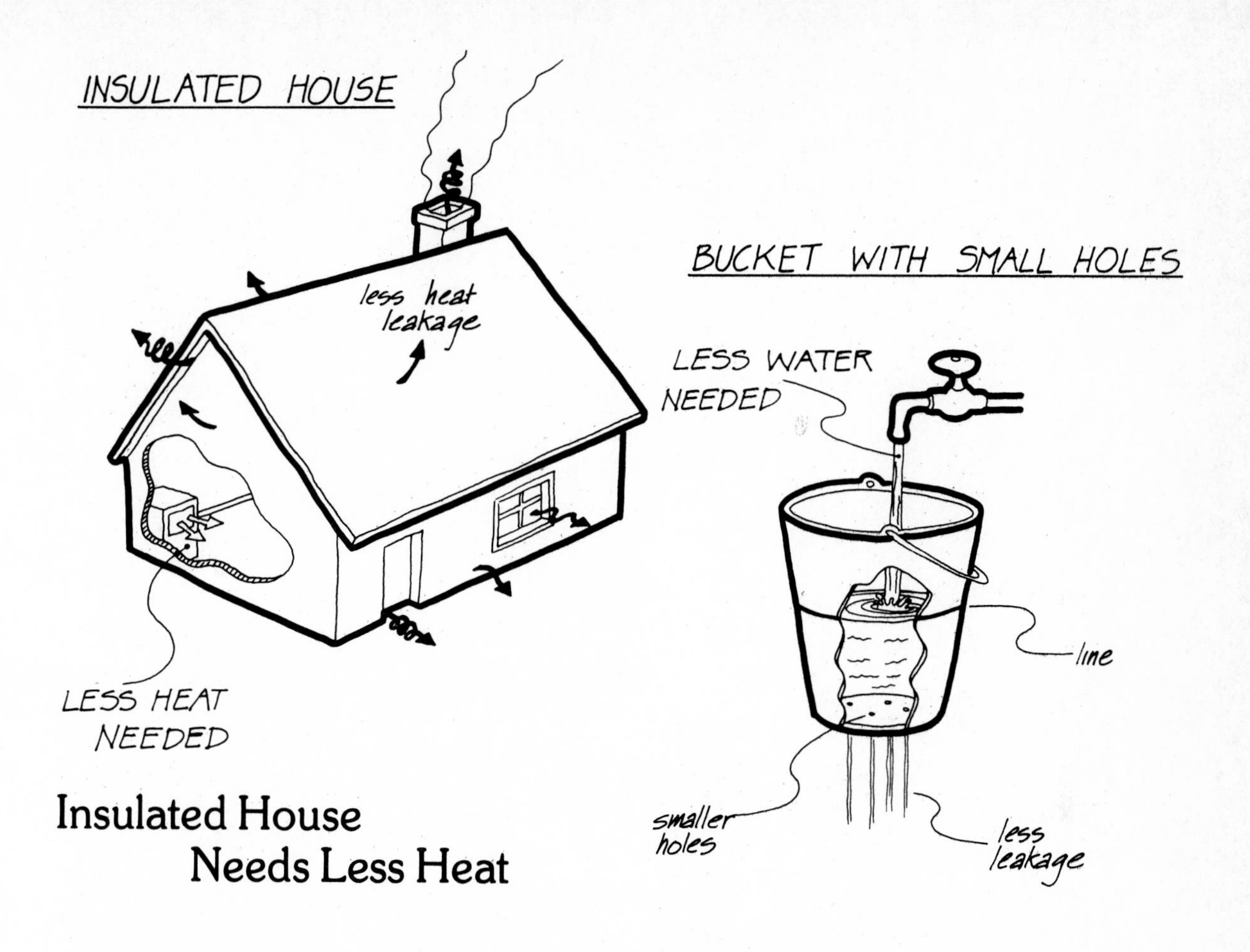

Insulated House Needs Less Heat

the holes in the bucket are smaller, less water leaks out through them and less water is needed from the faucet.

Conserving heat in a house by better insulating is probably *more* important than solar heating the same house. A solar heating system that could save half of your heating bill might cost several times more than better insulation, which could also save half of your heating bill. If your house is already well insulated, *then* solar heating is probably worth the expense.

SOLAR RADIATION 9

Now you've seen how houses lose heat. But how can the sun help to provide house heat? Can the same kind of system be used for hot-water heating or swimming-pool heating? To answer these questions, first we have to know something about solar radiation.

Remember that radiation heat flow (already discussed in Chapter 5) is a form of heat flow that needs no medium—such as air, water, or metal—to flow through. Solar radiation is like radiation heat flow, but it differs in important ways from the radiation heat flow already discussed. The distinctions between the two kinds of radiation are critical ones, especially with regard to solar heating.

Solar radiation and radiation heat flow belong to a class of radiation called *electromagnetic* radiation. X-rays, light waves, microwaves, television waves, and radio waves also belong to this class. Sound is another class of radiation. Sound radiation is subdivided into various pitches. Kettledrums and tubas are designed to handle low pitches, whereas flutes and violins can handle high pitches. Even different species of animals have ears that can hear different pitches. A dog's ear or a bat's ear can hear high pitches that human beings can't hear. Solar radiation and radiation heat flow differ in the same way as sounds of different pitch do. In fact, you could think of solar radiation as *high-pitched* radiation heat flow.

In terms of solar heating, radiation heat flow and solar radiation differ in two important ways. First, color is a very important factor in solar radiation but is not too important in radiation heat flow. If a surface is exposed to the sun it will get very hot if it's painted black or a dark color, but it won't get very hot if it's painted white or a light color. Benjamin Franklin reportedly discovered this fact by laying a black cloth and a white

Radiation Comes in Different Pitches

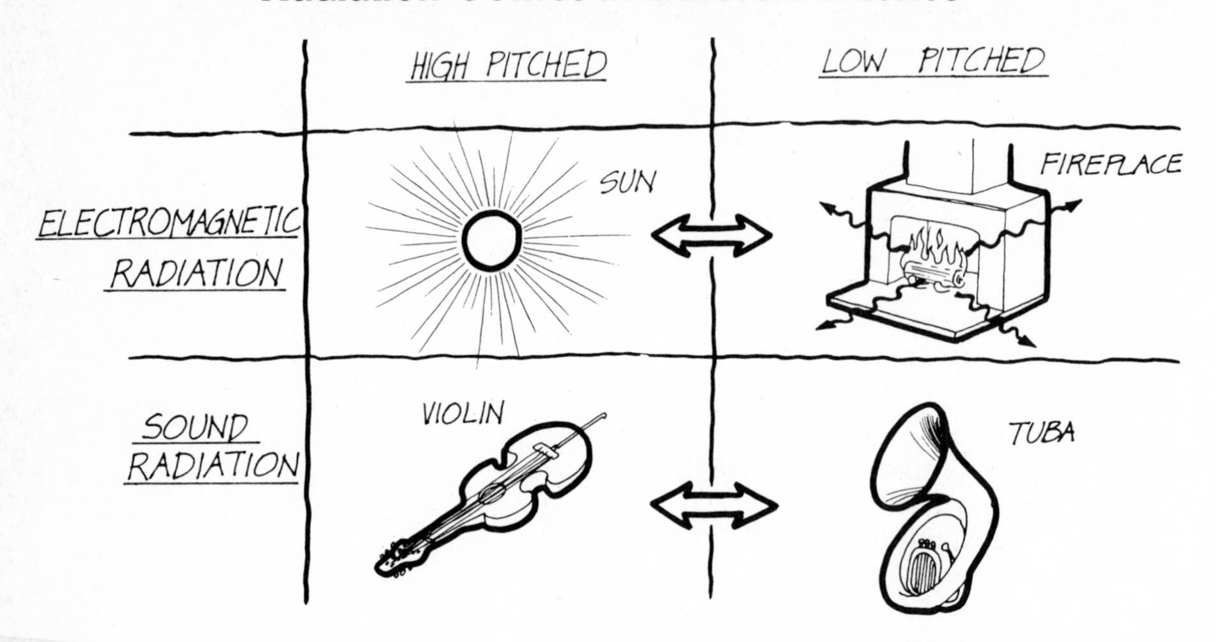

cloth on the snow on a sunny winter day. The snow melted much more quickly under the black cloth than under the white one because the black one absorbed more solar radiation. The white cloth reflected the sunlight away, so it didn't absorb as much solar radiation and didn't melt as much snow. We will use a hollow arrow ( ) to denote solar radiation to distinguish it from the other kinds of heat flow.

Solar Radiation is Sensitive to Color

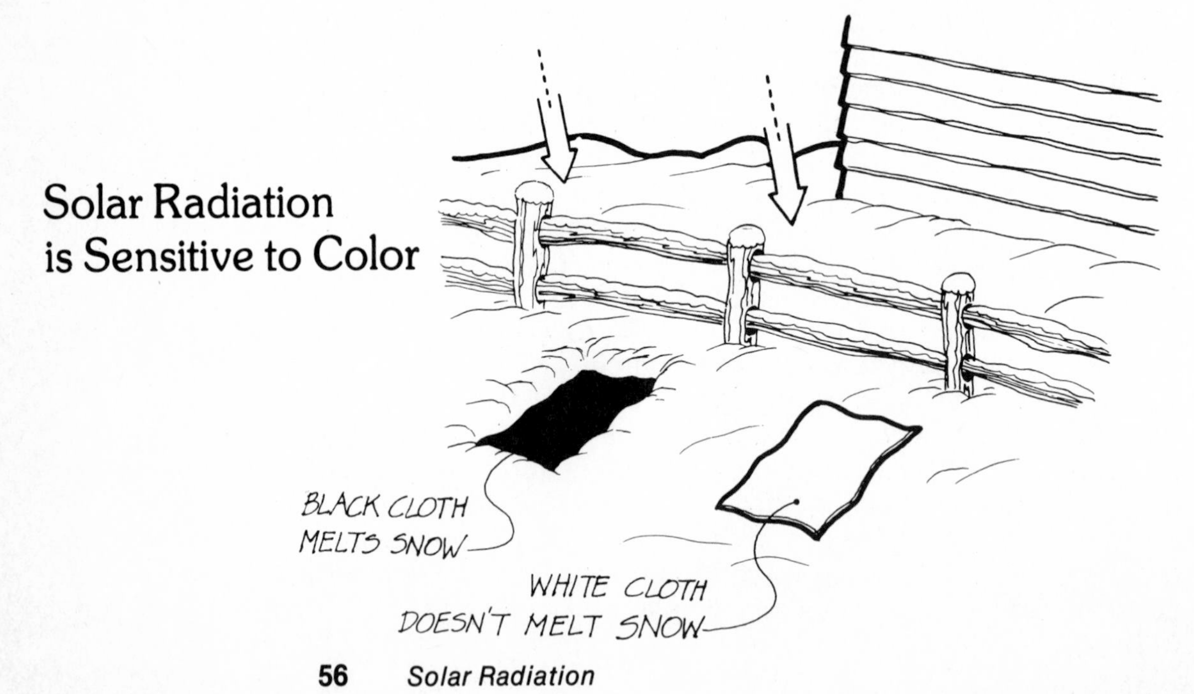

Unlike solar radiation, radiation heat flow reacts in the same way to different colors. If we held a black cloth and a white cloth in front of a fireplace, both would get just as hot.

Radiation Heat Flow is not Sensitive to Color

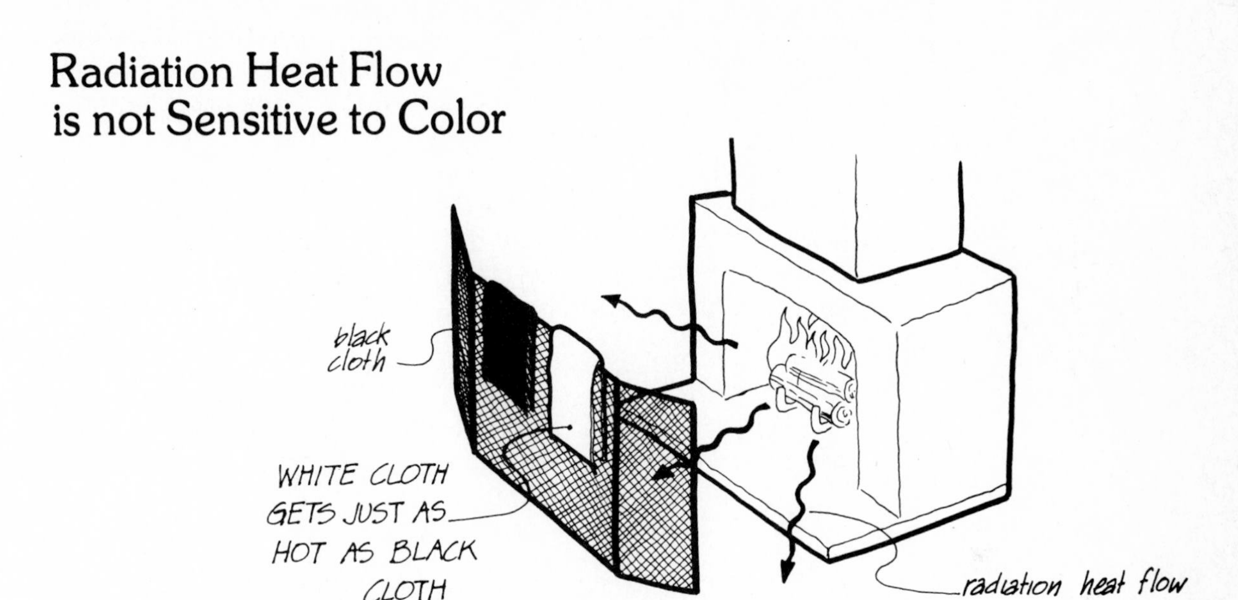

Recall that with radiation heat flow, the only way we could reflect the radiation was with a shiny surface. But with solar radiation both shiny surfaces and white or light-colored surfaces reflect the radiation. For example, you can get a worse sunburn on the beach than you can get on a golf course. On the beach you are exposed to solar radiation both directly from the sun and indirectly from the sun reflecting from the sand, but on a golf course lawn, you get the radiation only directly from the sun.

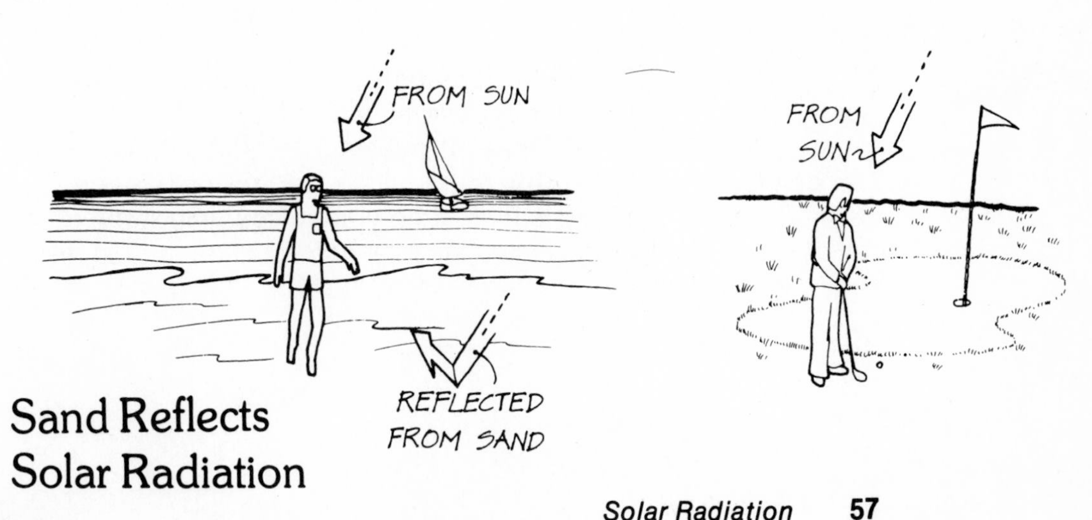

Sand Reflects Solar Radiation

Solar radiation and radiation heat flow differ in a second important way. Solar radiation can go through transparent materials like glass that radiation heat flow cannot go through. If you held a piece of glass in front of you, it would block the radiation heat flow from a fireplace; but it wouldn't stop solar radiation from coming through. The tendency for solar radiation, but not radiation heat flow, to pass through glass is sometimes called the *greenhouse effect*. Heat can enter a greenhouse through its glass roof, but the same glass acts like a solid wall to heat trying to leave the greenhouse by means of radiation. Plastic materials behave a little differently. Although both glass and plastic let solar radiation pass through them, some plastic materials also let radiation heat flow through.

Solar Radiation Goes Through Glass

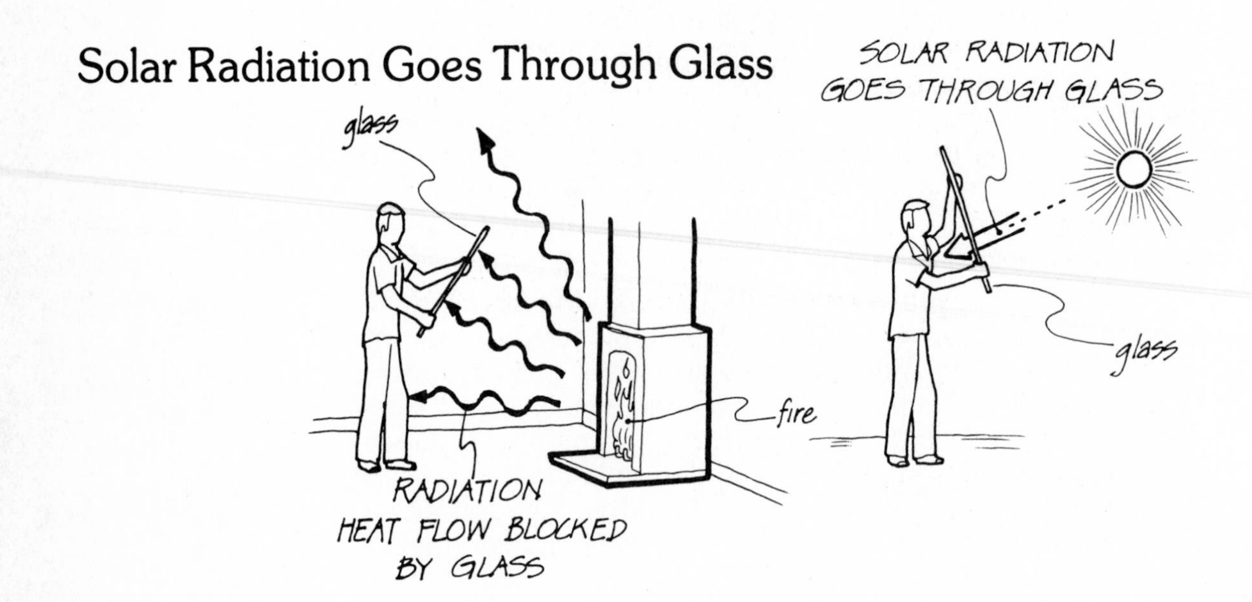

Other than the solar radiation's color sensitivity and its ability to pass through transparent materials, it behaves very much like radiation heat flow. It travels in straight lines, and it can be blocked by even thin material. To get out of the sun you get in the shade—the shade is simply anyplace where the solar radiation has been blocked. A tree can provide shade even though its leaves are very thin.

SOLAR COLLECTOR ORIENTATION 10

A solar heating system has two major parts: the solar collectors and the heat storage. The collectors capture the solar radiation that strikes them. As much as possible of this heat is then stored in the heat storage, and from there it is delivered to the house.

Solar collectors can be thought of as big nets which gather the solar radiation that strikes them. Just as a bucket in a rainstorm collects rain, so a solar collector collects the solar radiation raining down on it.

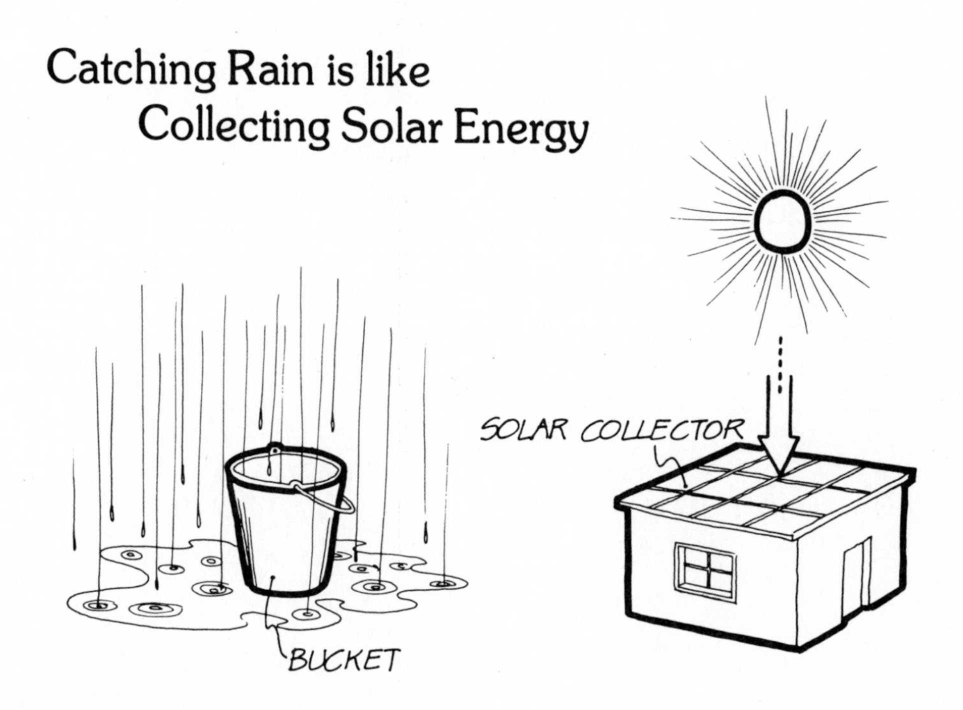

If you were trying to collect the most rain you could in a bucket, you'd hold the bucket at the proper angle to catch the rain. The open top of the bucket is the rain collector, and you'd point it toward the rain. If the rain were coming straight down you'd hold the top level, but if the rain were driven by the wind you'd tilt the bucket so that the top would still face the rain. Similarly, a solar collector should point toward the sun to pick up the most solar radiation possible. How can you tell if the sun is pointed at the collectors? If a small peg sticking straight out from the collector leaves no shadow, then the collector is pointed straight at the sun.

A Solar Collector Should Face the Sun

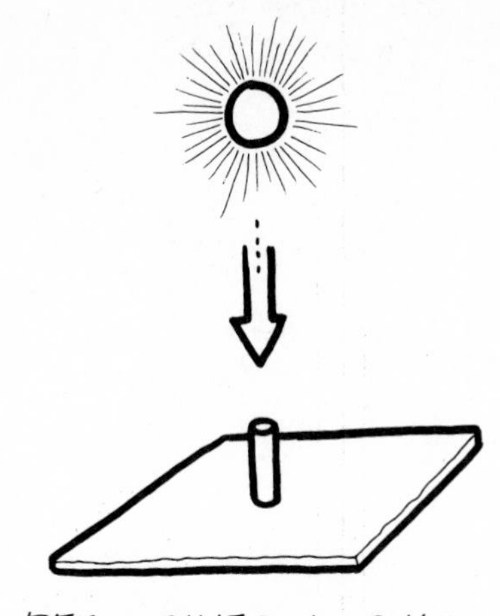

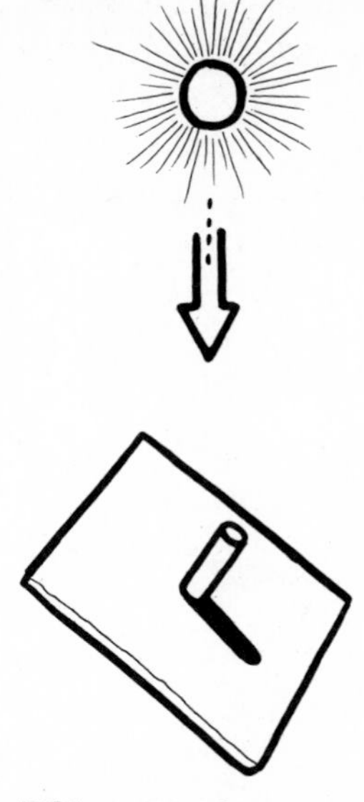

Collector Facing the Sun Leaves no Shadow

The sun's position in the sky changes during the *day* as the earth rotates. In the morning the sun rises in the east; in the evening it sets in the west. At midday the sun is highest in the sky, but it is never directly overhead (at least not in the United States). If you were to face directly south for a whole day the sun would rise at your left, travel in an arc through the sky, and set at your right.

The sun's position also changes with the *season.* In the winter, the sun is lower in the sky at midday than it is in the summer. You can get a better idea of the seasonal change of the sun's position by thinking in terms of shadows. Since the sun is higher in the sky in the summer, your noontime shadow will be shorter than it will be in the winter.

The sun's position in the sky changes from hour to hour and also with the season, yet we want to catch the most sunlight possible from a flat-plate collector that can't move so that it always faces the sun. To do this, we compromise so that the collector is pointed south, whereby it most nearly faces the sun during the midday hours, even though some solar radiation is lost in the morning and evening hours. Although there is a best direction to face a flat-plate collector, depending on the local latitude (less tilt is needed in the south than the north), a tilted collector that generally faces the south performs adequately.

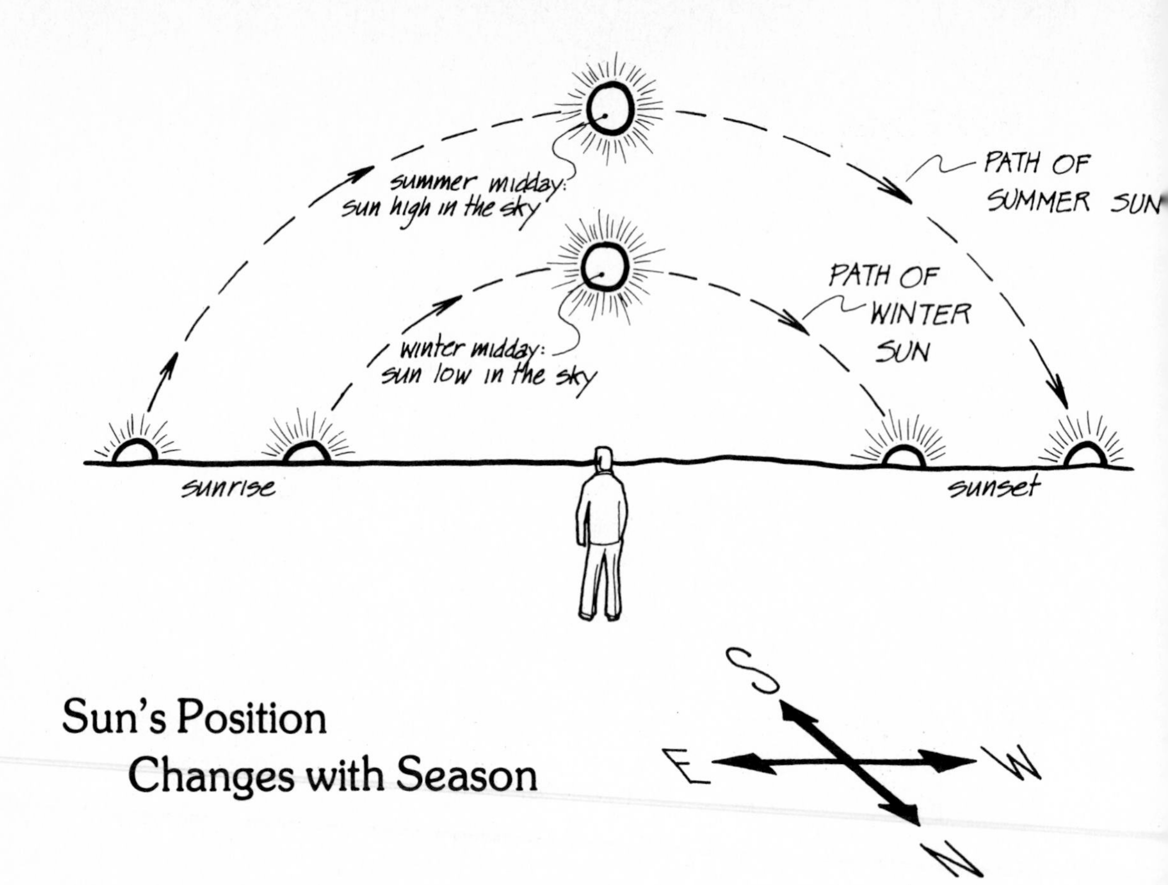

Sun's Position Changes with Season

Noon Sun is Higher in Summer Sky

Solar Collectors
Should Tilt to the South

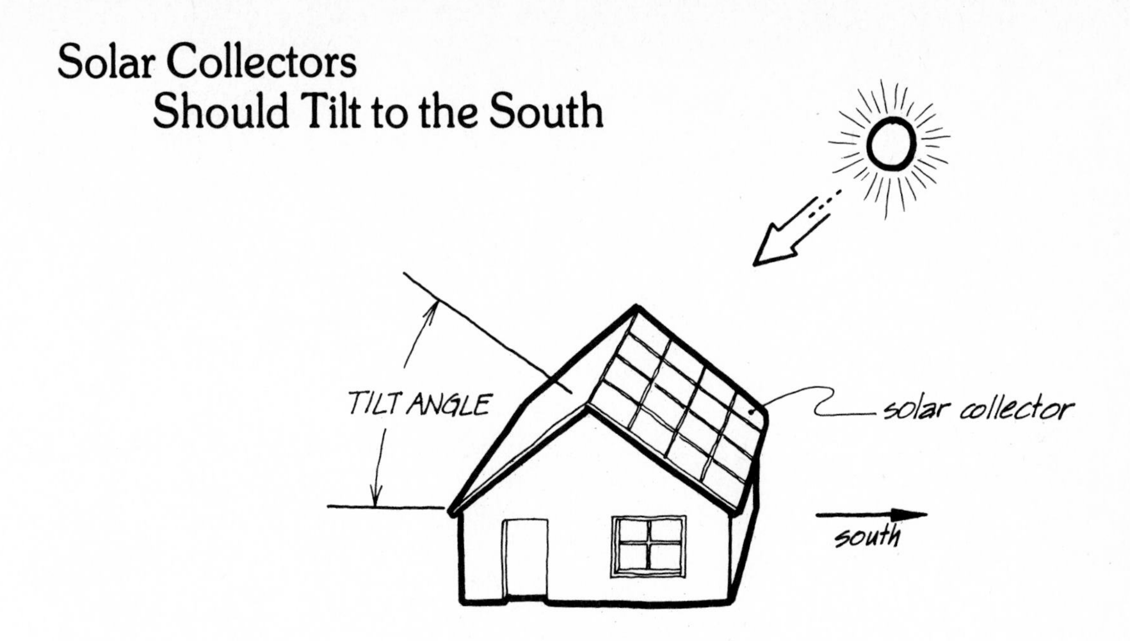

House-heating systems usually have a comparatively high tilt angle, since houses need the most solar heat in the winter when the sun is low. Hot-water systems need solar heat in both summer and winter, so they have a medium tilt angle—a compromise between facing the sun in the summer and the winter. Solar swimming-pool heaters usually have the lowest tilt angle. They need heat mostly in late spring, summer, and early fall when the sun is highest in the sky.

Tilt Angle
Depends on Application

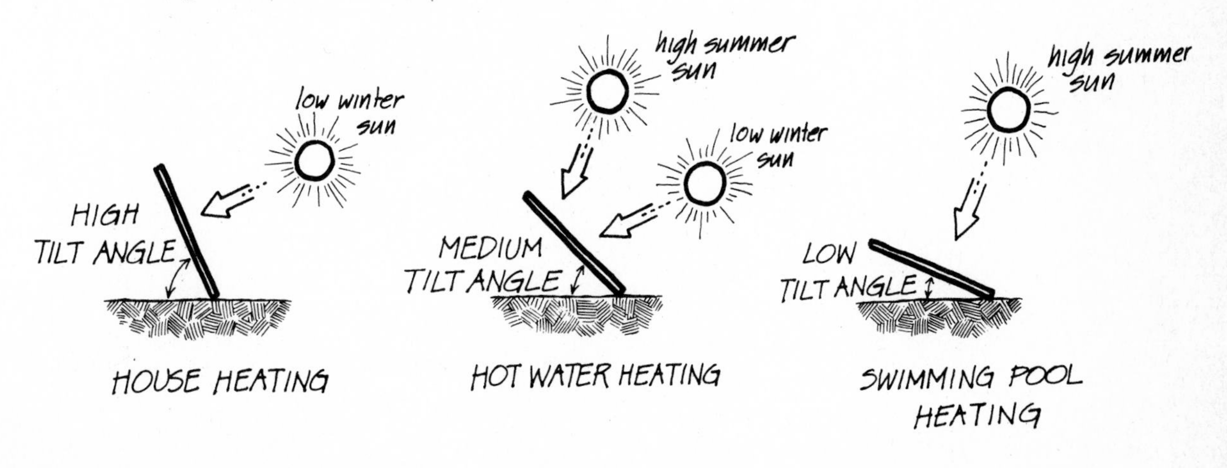

A SIMPLE SOLAR COLLECTOR 11

Now we need to learn what a solar collector is. The simplest kind of flat-plate solar collector is just a black-painted metal plate with a layer of insulation on the back side. It's painted black because black best absorbs solar radiation, as we discussed in Chapter 9. Since insulators prevent heat from conducting through them, the insulation on the collector prevents heat loss from the collector's back side.

If we faced the plate toward the sun at noon, it would begin to get hot as it absorbed solar heat. How hot would it get?

A Simple Solar Collector

We can learn a lot about a solar collector by thinking of it in terms of a rainwater collector. Imagine that we are building a device to collect rainwater. As a start, we get a big tray to catch the rainwater, just as a solar collector "collects," or absorbs, solar radiation. We attach a capped pipe to the tray so that all the rain falling in the tray will flow down the pipe and be caught.

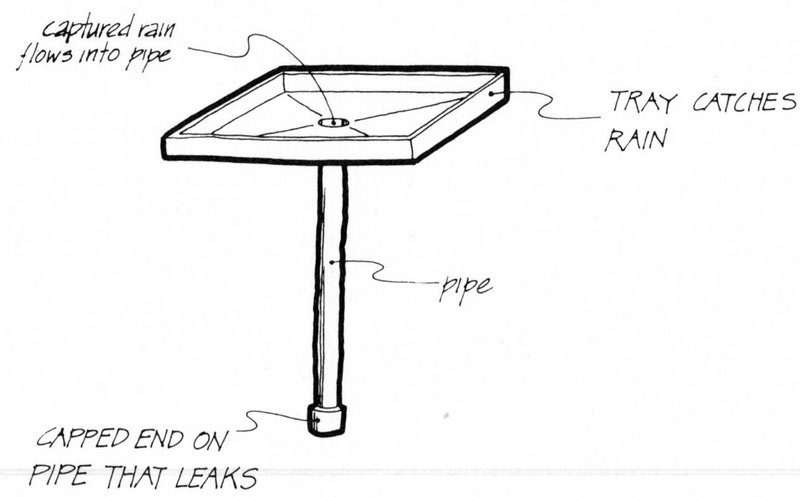

A Rainwater Collector with a Leaky Pipe

Suppose the only capped pipe we can find is one that leaks. As the tray collects rain, the pipe fills; but rainwater flows out through the leak. The deeper the water gets in the pipe, the more water leaks out. At some point the depth gets so high that the flow out the leak balances the incoming flow of rainwater. Once this point is reached, the water gets no deeper. When the incoming rainwater exactly balances the outgoing water loss through the leak, the depth of water in the pipe is at its *stagnation* depth.

Similarly, the stagnation temperature is how hot a collector gets when the incoming solar radiation exactly balances the heat losses that take place by radiation and convection. To continue our analogy, the tray captures rainwater as the solar collector captures solar radiation. Just as the water in the pipe gets deeper and deeper, so does the temperature of the plate become hotter and hotter. And just as more water flows out through the leak as the water in the pipe gets deeper, so more

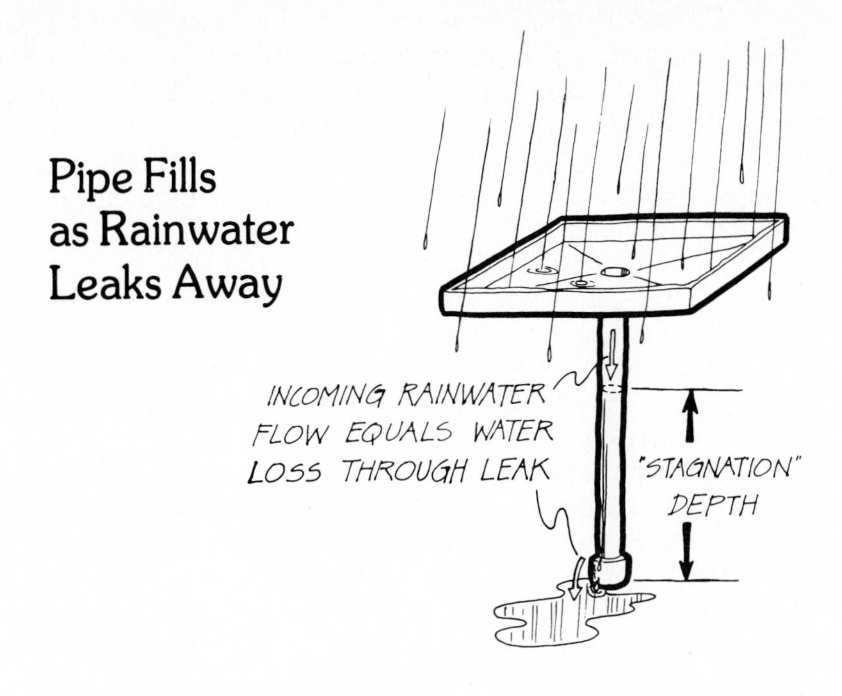

Pipe Fills as Rainwater Leaks Away

heat leaks off the front of the solar collector by radiation and convection as the plate gets hotter. At some point the plate gets so hot that the heat losses exactly balance the incoming solar radiation. Once this point is reached, the plate gets no hotter.

Collector Gets Hot as Heat Leaks Away

The stagnation temperature of a surface—the hottest a surface exposed to the sun can get—depends on how much solar radiation is absorbed. The amount of radiation absorbed depends on the color of the surface, the angle of the sun's rays striking the collector, and how much sunlight actually strikes it. The best setup to use to absorb solar radiation is the one just described: a black surface squarely facing the sun on a clear day. A light-colored surface, a surface not directly facing the sun, or an overcast day will reduce the amount of solar radiation absorbed by the surface.

How will less absorbed radiation affect the stagnation temperature? Let's use the analogy of the rainwater collector again to understand what happens. Suppose a light rain is falling on the tray rather than a heavy rain—that's comparable to less solar radiation striking the solar collector. With less rain falling, the tray collects less and the incoming rainwater flow is less. Recall that when we discussed house heating, we found that a lower water depth in the bucket resulted in less flow out through the bucket's holes: high depths are associated with high leakage; low depths, with low leakage. Thus, if less rainwater is flowing down the pipe, the depth needn't be as high to force all of the incoming water out through the holes. The water in the pipe wouldn't be as deep before the flow out through the

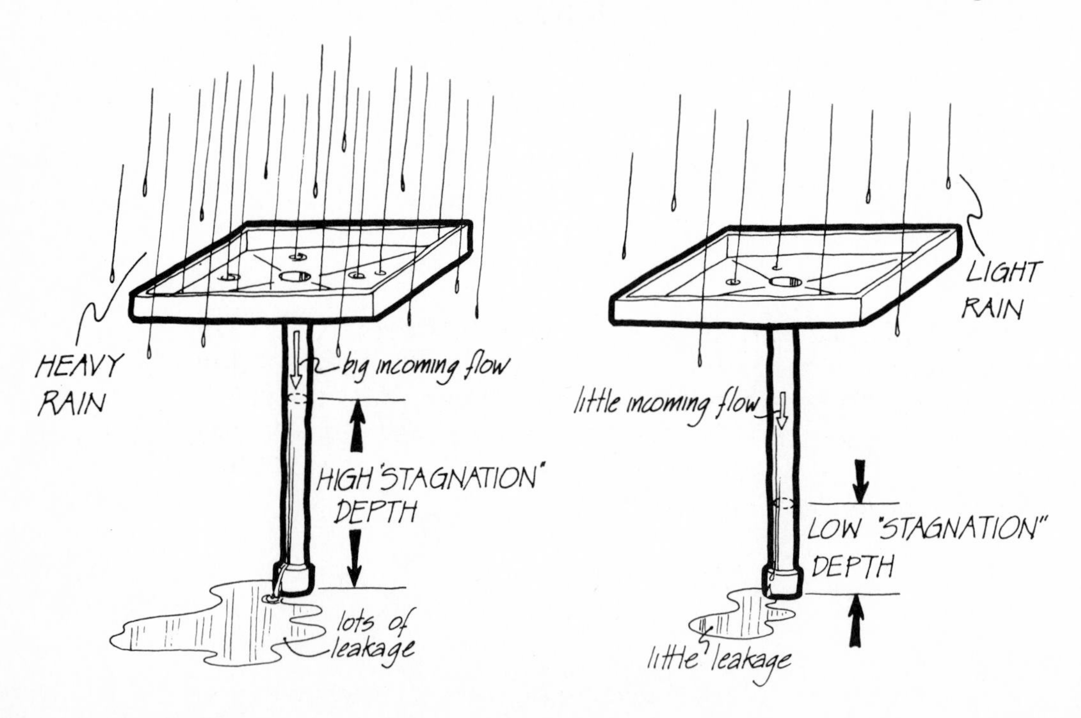

Pipe Fills Higher with Heavy Rain

holes balanced the incoming rainwater flow. We can conclude from this that if the incoming rainwater is less, the stagnation depth also goes down.

How can we apply this analogy to the solar collector? We would expect the stagnation temperature to be less if the absorbed solar radiation was less (since the stagnation depth was less when the rainfall was less). When less solar radiation is absorbed, our simple solar collector starts to get hot. It will lose more and more heat as it gets hotter. Eventually the heat losses will equal the amount of solar radiation absorbed. But since the solar radiation absorbed is less, the collector won't be very hot when the losses balance the absorbed radiation.

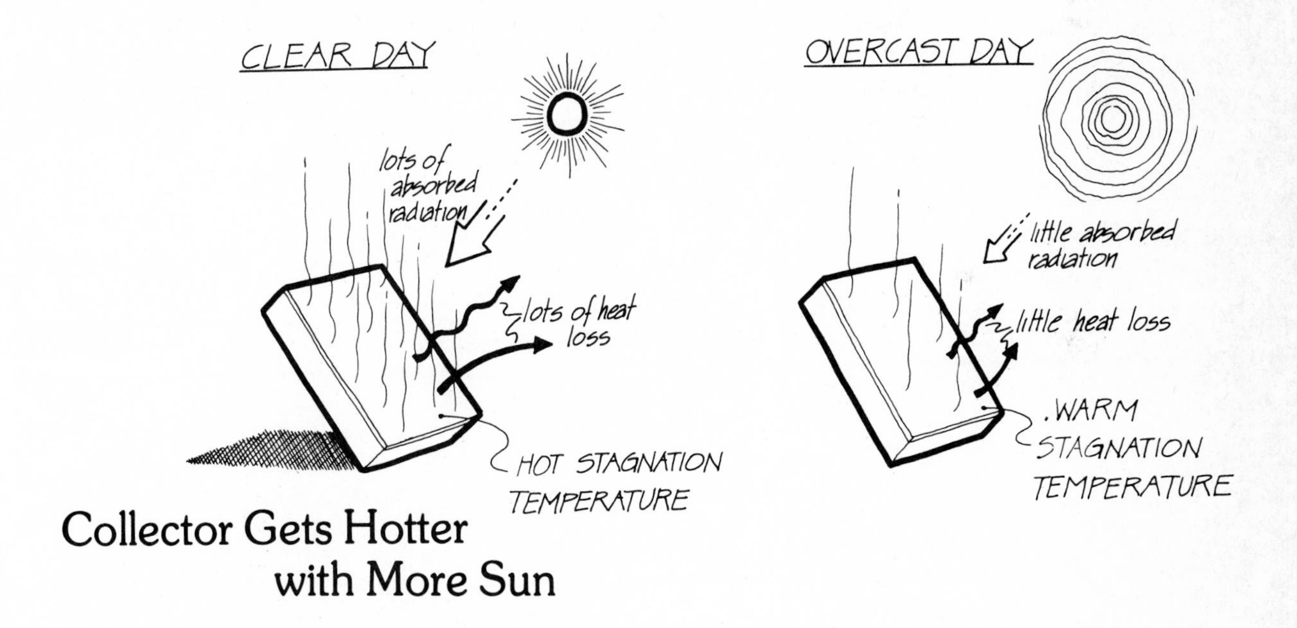

Collector Gets Hotter with More Sun

How hot our simple solar collector gets depends on how much solar radiation it absorbs. We've gotten the plate hot, but we haven't gotten any heat from it yet—the plate may be collecting solar radiation, but it's not doing us any good. You can't pay your heating bills with a hot black plate. We want *heat* from a solar heating system, not temperature.

A SOLAR HEATING SYSTEM 12

Not only do we want to *extract* heat from our collector, but we also want to *store* this same heat to use after the sun goes down. We can accomplish both purposes by using transport heat flow. Our simple *insulated plate* collector can be modified by soldering metal tubing onto the back of the plate and pumping water through the tubing.

Extracting Solar Heat by Transport Heat Flow

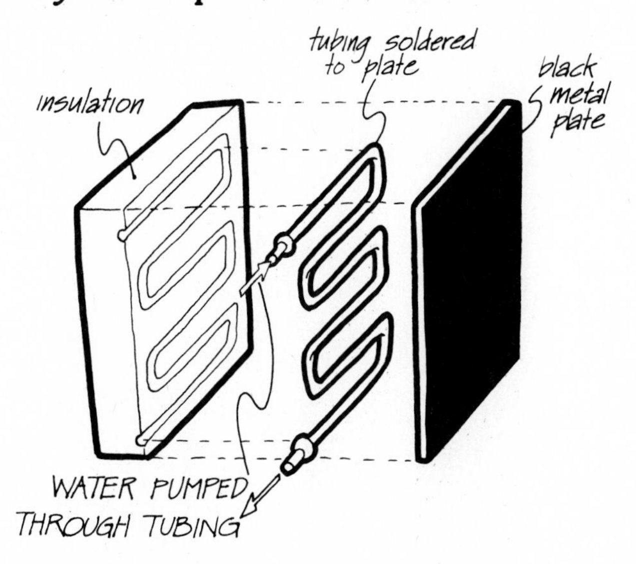

Using both metal tubing and metal plate is important, since heat flows easily through metal by conduction. In this case, the heat flows from where the sun hits the plate to where the tubing is soldered to the plate. Convection into a liquid, like water, is excellent, so the heat flowing into the tubing easily heats the water. Insulation is again used on the back side of the collector to prevent heat from leaking away.

Solar Heat is Conducted through Metal Collector

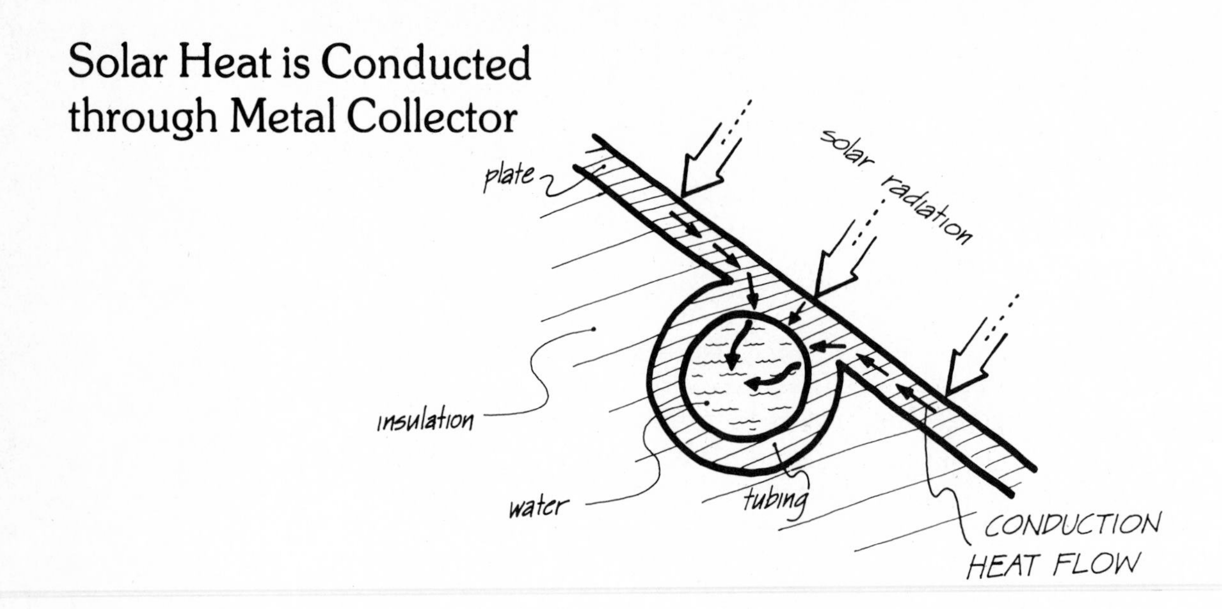

After solar heat flows into the water, it can be transported from the collector by pumping the heated water to a heat storage tank. A heat storage tank is a water-filled tank covered with insulation to prevent heat loss. The pump transports water from the storage tank to the collector, where it is solar heated, and is then pumped back to the storage tank in a continuous cycle. We

Solar Heat is Transported to Storage Tank

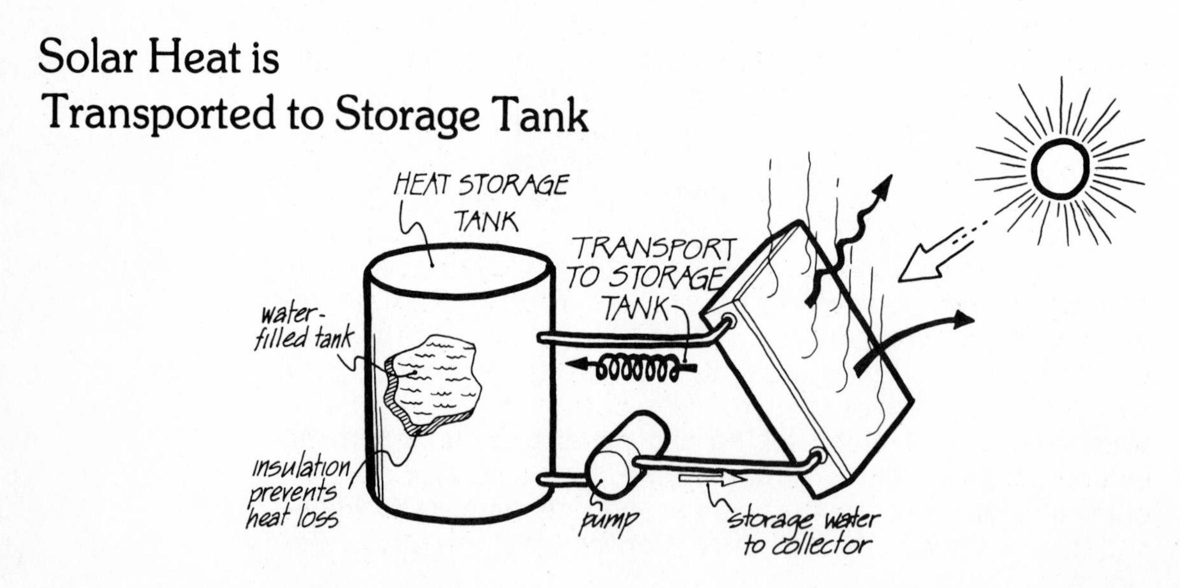

learned earlier that transport heat flow depends on how fast the water is pumped. Let's assume that we've "sized" the pump so that the water flows fast enough to let heat flow easily out of the collector and into storage. At the same time, the pump won't be so big that the power needed to run it will be significant compared with the heat collected by our system.

To return to the water analogy: we don't want only a high depth in the pipe of the rainwater tray. We want to extract and store some rainwater for use when the rain stops. If we use a fat tube to connect the bottom of the pipe to a big tank, some of the rainwater will flow into the tank. The tube is fat enough so the water in the pipe can flow easily into the tank.

A Rainwater Collector with a Storage Tank

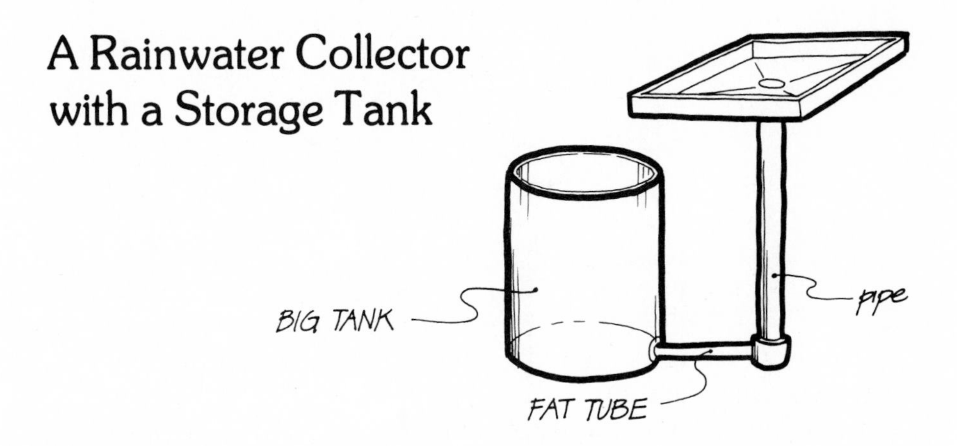

The fat tube is analogous to the pump used in the solar heating system. Just as the pump lets transport heat flow easily from the solar collector to the heat storage tank, so the fat tube lets volume flow easily from the rainwater collector to the rainwater storage tank.

Suppose a heavy rain is falling on the rainwater tray. The water will flow down the pipe and start filling the tank. Some of the incoming rainwater will flow into the tank, and some will flow out the leak. Since we want to store as much rainwater as we can, we'd like as little as possible to leak away. When only a small fraction of the incoming rainwater leaks away, we say we have high *collection efficiency*. We're efficiently using the rainwater by storing lots of it and letting only a little leak away. By contrast, we have low collection efficiency when a large fraction of the incoming rainwater leaks away and only a small fraction is stored.

When do we get high collection efficiency and when do we get low collection efficiency? The easiest way to understand collection efficiency is to remember that ***high*** efficiency happens only when leakage is *low*. Leakage is low only when the depth of water in the pipe is low—a nearly empty pipe leaks less than a full one.

High Efficiency Collection with Empty Pipe

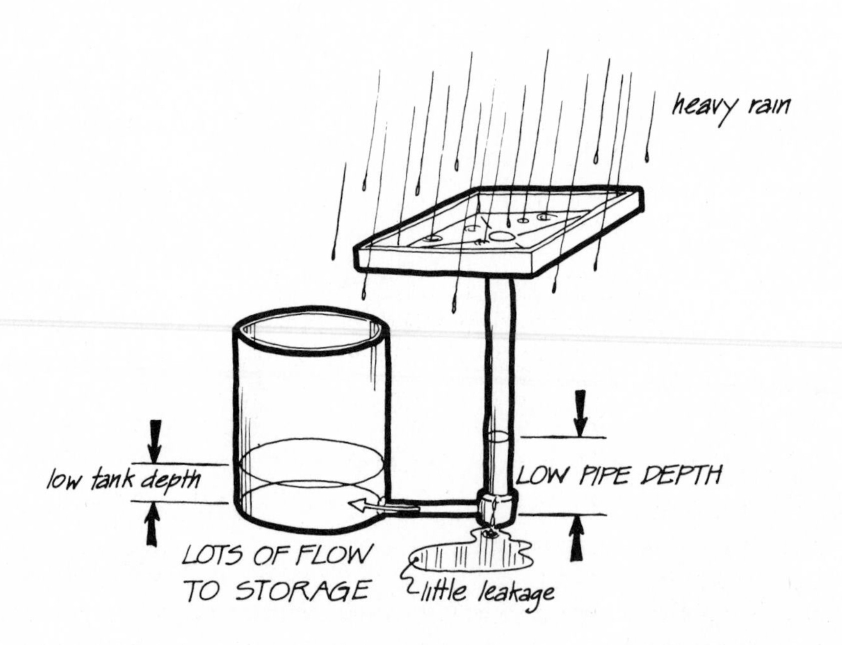

In the rainwater collecting apparatus we're discussing, we could get high efficiency *or* low efficiency, depending on the depth of the water in the pipe. For example, if the water depth in the tank were low, the depth of water in the pipe would also be low (the fat tube makes sure both have about the same depth). A low pipe depth means high efficiency: little leakage and lots of flow to storage. The same apparatus would give us low collection efficiency if the storage tank were nearly full. A full tank would indicate that the water depth in the pipe was also high (the fat tube insures this), resulting in high losses. Of course, if most of what is coming into the pipe were lost through leakage, the collection efficiency would be low.

Low Collection Efficiency with Full Pipe

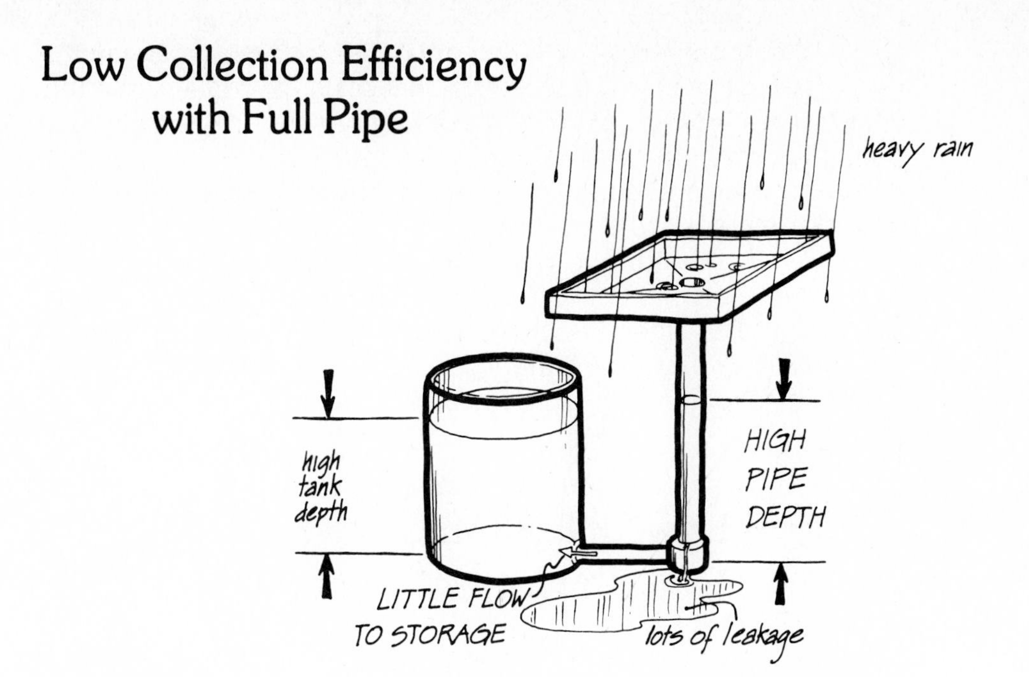

Now that you have some understanding of collection efficiency, lets's return to the solar heater. We'll assume it's about noon on a bright, sunny day: the collector is absorbing lots of solar radiation.

When the heat storage tank is cool, then cool water is pumped through the collector. The sun warms it somewhat, so it comes out a little warmer than it went in, but on the average the

High Collection Efficiency with Cool Collector

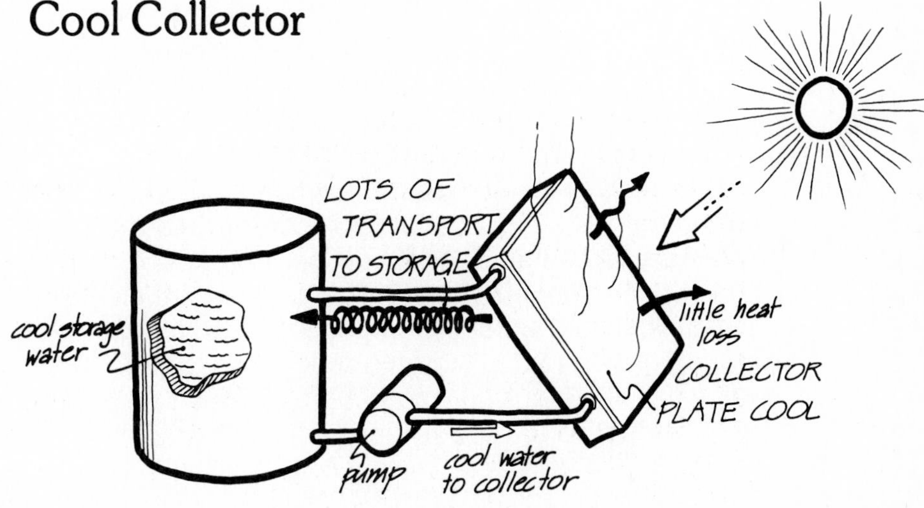

solar collector plate stays fairly cool. With a cool solar collector plate, heat losses by radiation and convection are small, so little heat is lost. Lots of heat goes to storage and efficiency is high. Similarly, when the rainwater tank has little water in it, the depth in the pipe is low, so little rainwater is lost through the leak. Lots of water goes to storage and efficiency is high.

When the storage temperature is hot, the water being pumped to the solar collector is also hot. The collector, in turn, gets hotter. But when the collector gets hot, it loses more heat from its front plate by radiation and convection. Since more solar heat is lost, less heat is stored and the efficiency is low.

Low Collection Efficiency with Hot Collector

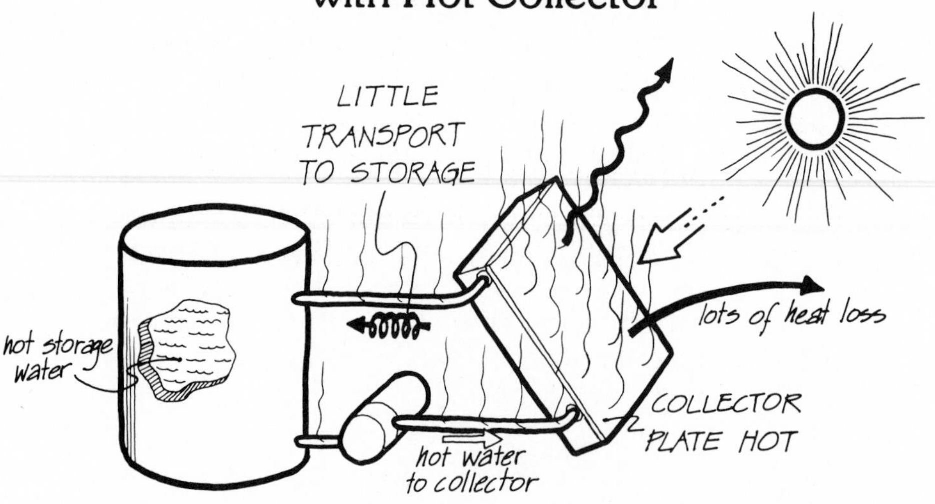

A hot collector means low-efficiency solar collection. We learned earlier that the hottest a collector can get is its stagnation temperature. But at its stagnation temperature its efficiency is precisely zero—it loses all the solar radiation it absorbs. If the water pumped to the collector is cooler than the stagnation temperature, some solar heat can be stored: the cooler the water, the higher the efficiency. The highest efficiency is achieved when very cold water is pumped into the collector. The tendency for efficiency to depend on temperature is an inherent feature of all solar heaters. It has an important effect on how solar energy can be applied to various kinds of heating needs.

SOLAR HEATING APPLICATIONS 13

Efficiency, as we've just learned, depends on the collector's temperature. A collector that stays cool loses little heat to the outdoors; one that's hot leaks a lot of heat. Consider a solar swimming-pool heater. Collectors on the roof, quite similar to the ones just described in Chapter 12, absorb solar radiation and heat the water being pumped through them. But, instead of using a storage tank, the pool itself stores the solar-heated water.

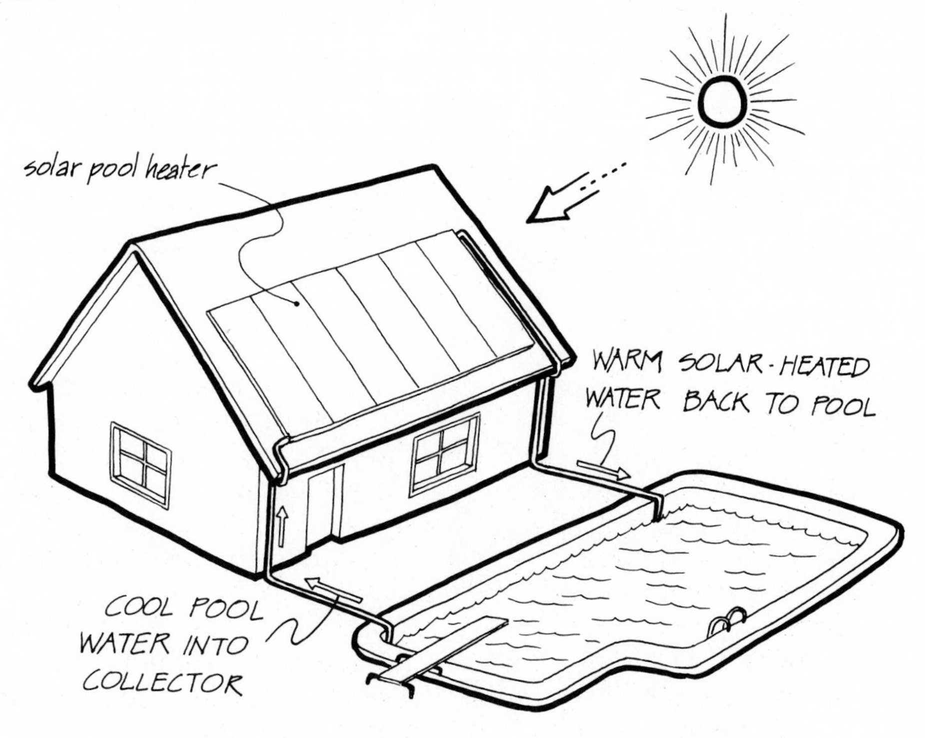

Swimming Pool Stores Solar Heat

A swimming-pool solar collector gets very high efficiency because it uses heat only at relatively low temperatures. The water pumped into the collector is cool water from the pool. The water going into the collectors might be 80° F and the water coming out might be 90° F, so on the average the collector stays fairly cool. Solar pool heaters are used mostly in the late spring and early fall, so the daytime temperature (when all the solar collecting is going on) is perhaps 60° to 80° F. Since heat loss is due to temperature difference, the small temperature difference (maybe 20° F) between the collector and the outdoors means little heat will be lost from a solar swimming-pool collector. Low heat loss, of course, means high efficiency. Even the simplest solar swimming-pool collectors can get efficiencies of 75 percent or more. Three-quarters of the solar radiation striking the collector ends up heating the pool water; only one-quarter of the heat is lost.

Swimming Pool Heater
Operates
at Cool Temperatures

For most other applications of solar heating—notably home heating and hot-water heating—conditions are not so favorable for high efficiency. For example, in solar heating your home, the temperature on a cold winter day might be 30° F outdoors at the same time the storage tank temperature is 130° F. A 100° F temperature difference between the collector and the outdoors means you would lose about five times the heat of a solar pool heater when the temperature difference was only 20° F. With losses this high, efficiency would be very low.

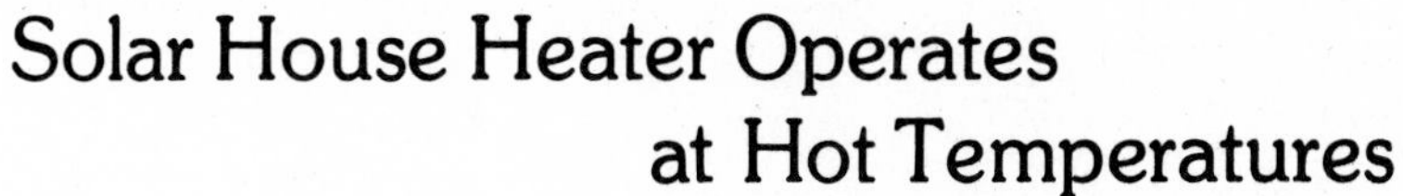

Solar House Heater Operates at Hot Temperatures

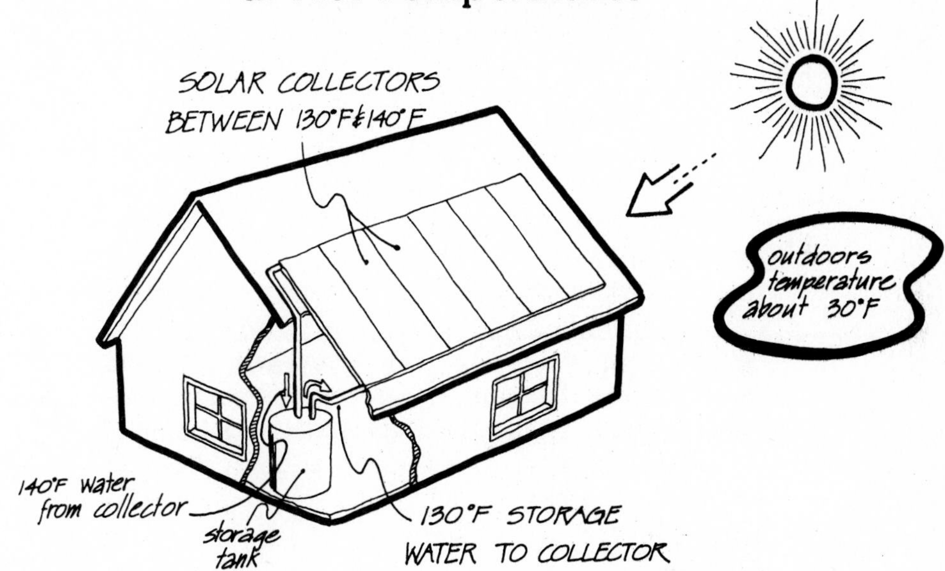

In heating water the problem is similar to house heating: it's cold outside and you want hot water. Lukewarm water coming from a solar water heater isn't good enough. As in house heating, the daytime temperature in winter might be about 30° F outdoors while the stored heat might be at 130° F. Again, a 100°F temperature difference between the collector and outdoors means that lots of heat is lost and efficiency is low.

Note that in the solar hot-water heater, the storage tank and the collectors are much smaller than those used for house heating. Since in most parts of the country more heat is needed to heat a house than to heat water for the same house, a solar hot-water heater is usually quite small compared with a solar *space-heating* system.

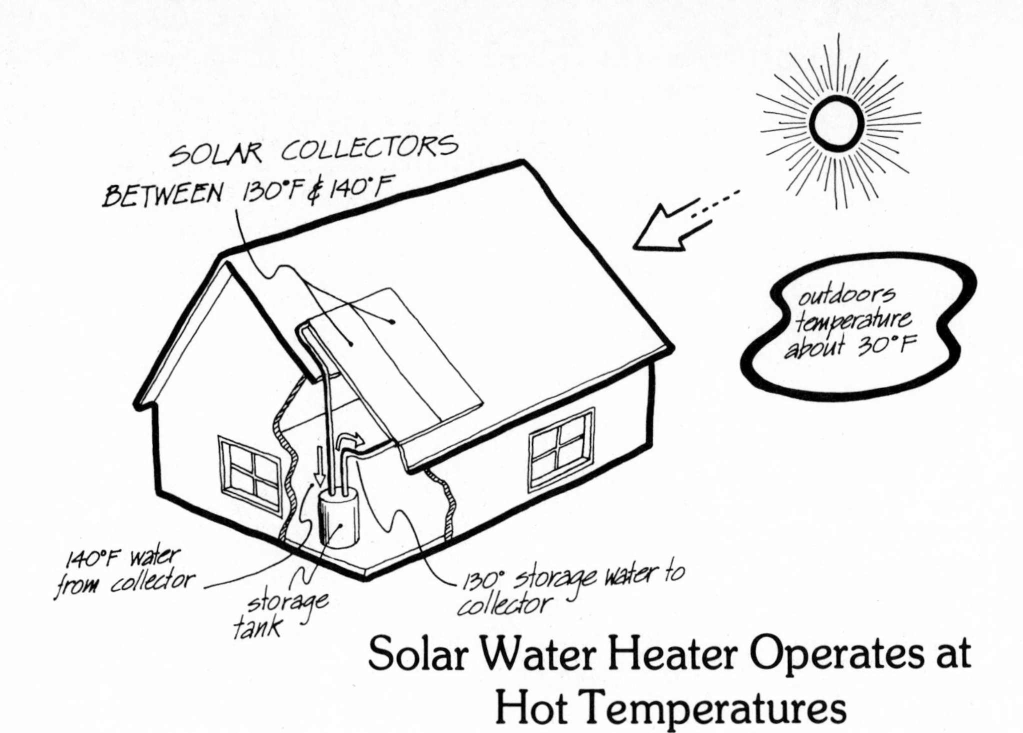

Solar Water Heater Operates at Hot Temperatures

For either house heating or hot-water heating a simple solar collector such as the one we've discussed so far won't work too well. What's appropriate for swimming-pool heating is too inefficient for house and hot-water heating.

GLAZING AND SELECTIVE SURFACES 14

How do we get both high temperature and high efficiency from a solar heating system? We need high temperature (130° F) for house and hot-water heating, but we also want high efficiency so the system will be economical. If we were only gathering rainwater, it would be easy to see how to get higher efficiency: we would try to stop the leakage.

If we simply tied a rag tightly around the leaking pipe, it would be harder for the water to leak out. With less leakage efficiency is higher, since a greater fraction of the incoming rainwater flows into the storage tank.

Rag Reduces Rainwater Leakage

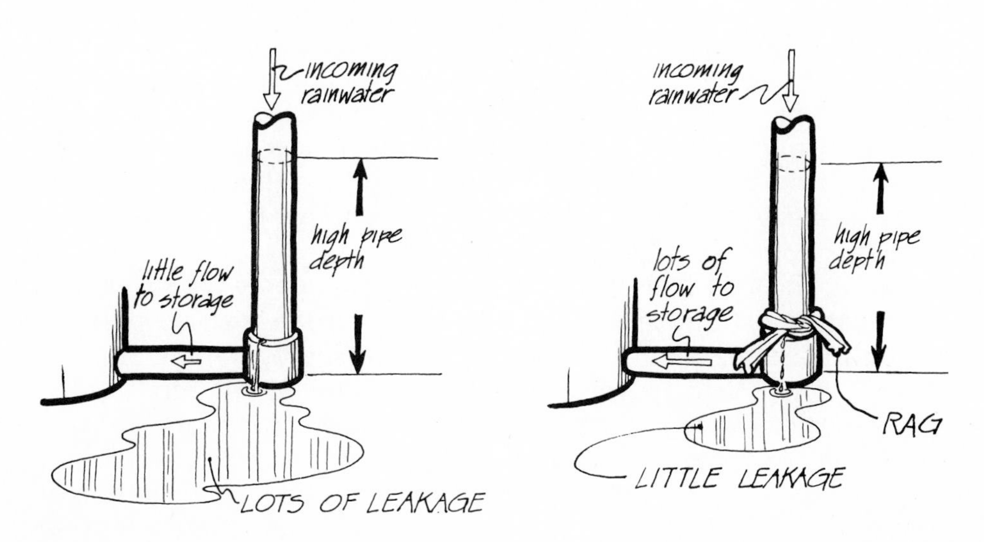

Remember that the water depth in the pipe in the rainwater collector is equivalent to the collector temperature in a solar collector. High rainwater collection efficiency is equivalent to high solar collection efficiency. If we want high solar efficiency *and* hot collector temperatures, we'll have to stop heat from leaking from the collector; just as we would use a rag to stop rainwater from leaking out of the pipe.

Glazing is one way to reduce heat loss from a collector. Glazing is a transparent cover, usually made of glass or plastic. It is held a few inches from the collector surface by the glazing frame. Since solar radiation passes through glass or plastic, almost all of it is still absorbed by the black collector surface. But with glazing, heat can't escape as easily.

Glazing is a Transparent Cover Over the Collector

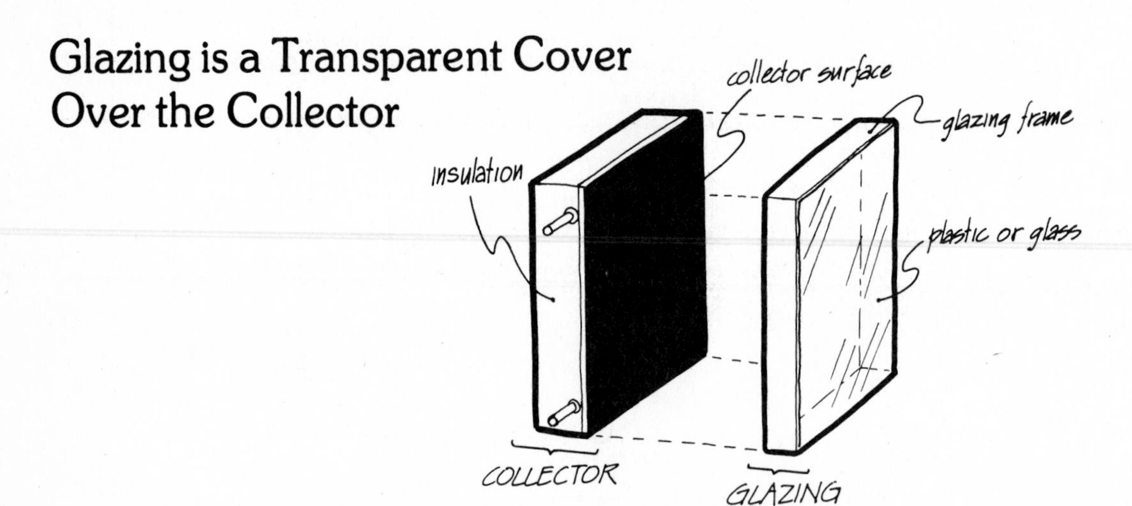

The biggest heat loss, by far, on an *unglazed* collector takes place by convection (although radiation losses can be one-fifth of the total). Convection losses are high mostly because the collector is exposed to the windy outdoors. As we have learned, when air blows over a surface, the surface loses heat easily. A collector on the roof of a house is seldom in still air.

Glazing prevents heat loss from the collector surface by isolating it from the wind. The air inside the glazing cover is "still" air; heat flows from the black surface by natural convection rather than by *forced* convection (the wind). First, some heat will leave the collector surface by means of natural convection; then it must flow into the inside surface of the glazing, again by natural convection. Next, conduction heat flow takes it through the glazing, and finally the wind blows it off the outer surface of the glazing by forced convection. You put pizza in a box for the same reason as a solar collector has glazing—the box prevents convection heat loss from the pizza.

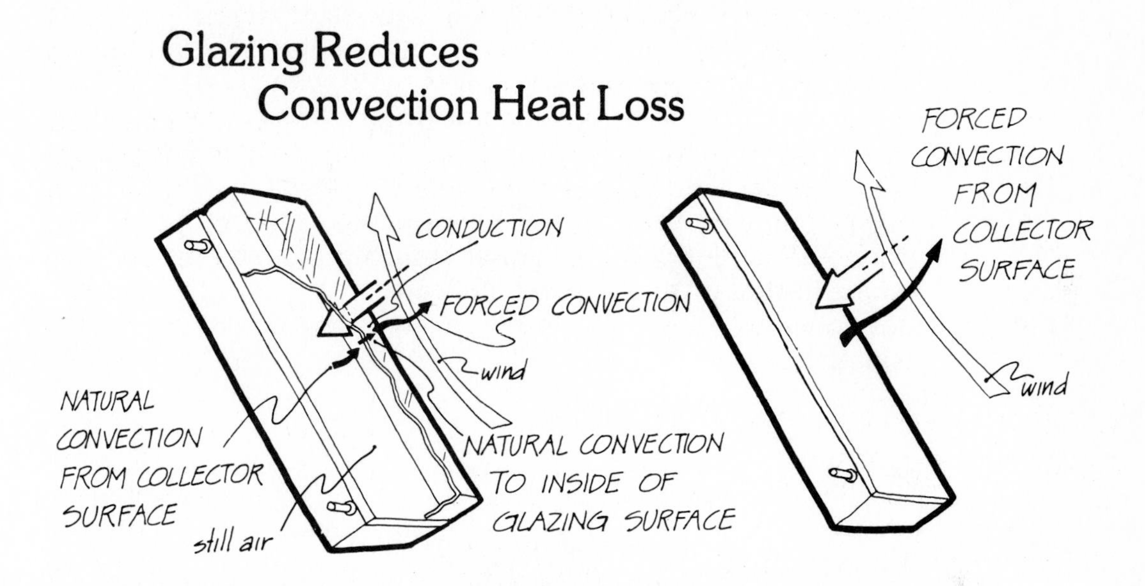

The glazing also reduces the amount of radiation heat flowing from the collector surface. Since radiation heat flow is mostly blocked by glass (and partially blocked by plastic), the radiation heat flow "sees" the inner surface of the glazing rather than the sky. The inner surface of the glazing is hotter than the "sky temperature," so radiation heat loss is less with the smaller temperature difference.

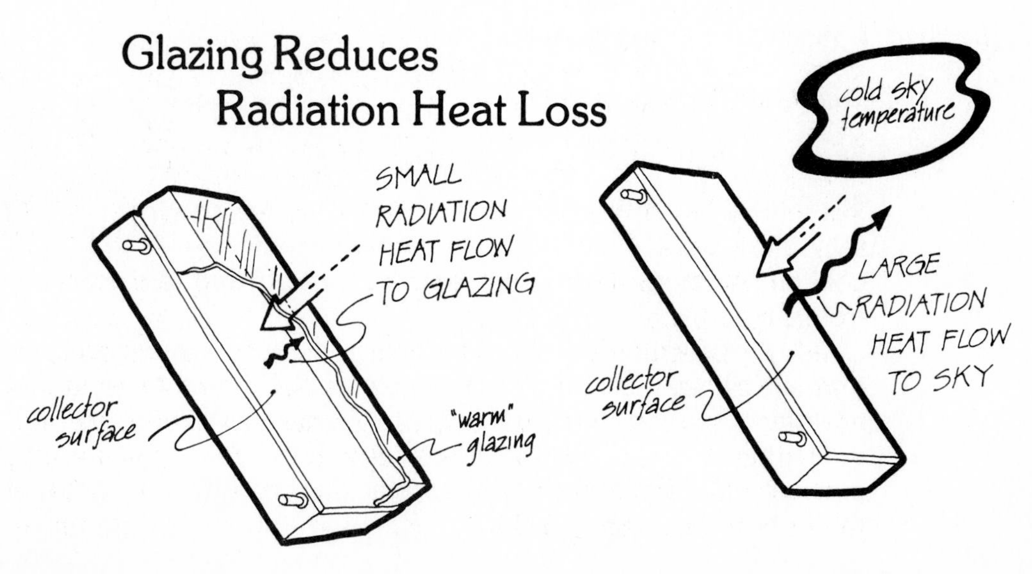

Another way to prevent heat loss from a collector is to coat its surface with a *selective* coating. The coating absorbs the high-pitched solar radiation just as black paint does, but it prevents most of the low-pitched radiation heat flow from leaving the collector surface. These special coatings are called selective because they react differently to solar radiation than to radiation heat flow. Usually they're used with glazing. A bare unglazed collector loses only one-fifth of its heat by radiation, whereas four-fifths of its heat is lost by wind convection. A glazed collector, though, has very low convection heat loss, so the radiation portion of heat loss becomes a substantial part of the total—about half. By further reducing the radiation portion of the heat loss from a glazed collector with a selective surface, the total heat loss of a glazed collector can be cut down substantially.

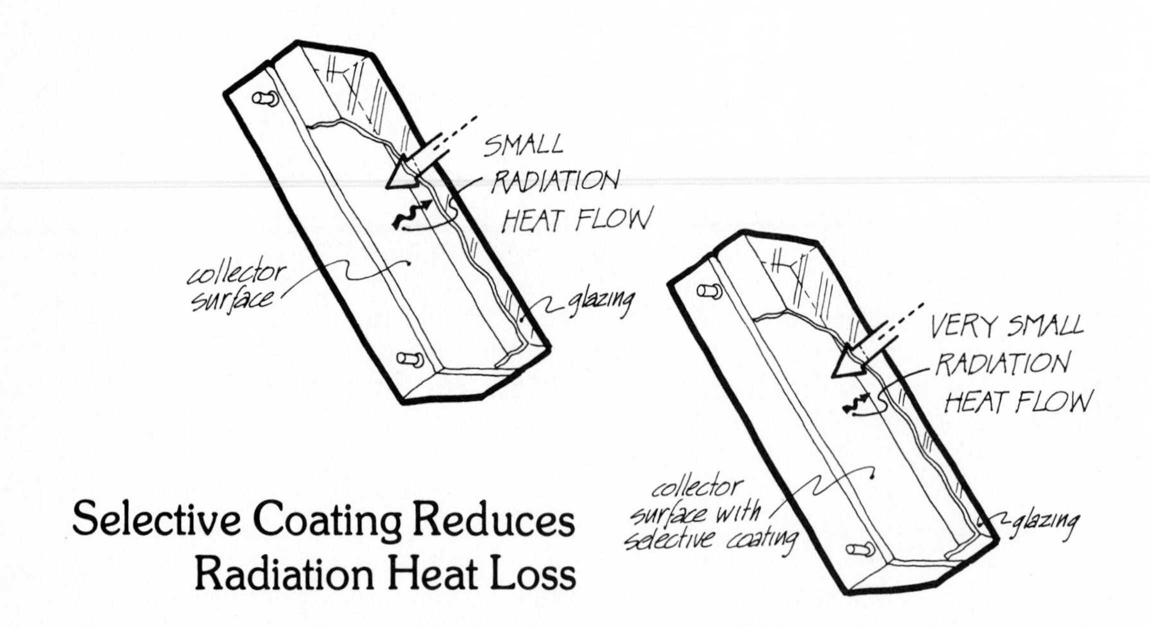

Selective Coating Reduces Radiation Heat Loss

The net effect of glazing and selective coatings is that they allow solar radiation in but prevent heat losses from the collector, just as the rag prevents rainwater from leaking away from a pipe.

A collector with glazing delivers high temperatures at high efficiency—exactly what is needed for house heating and hot-water heating. A glazed collector loses less heat to the outdoors than an unglazed one at the same temperature, just as the pipe with the rag leaks less rainwater than one without a rag at the same water depth.

Glazed Collector Loses Less Heat

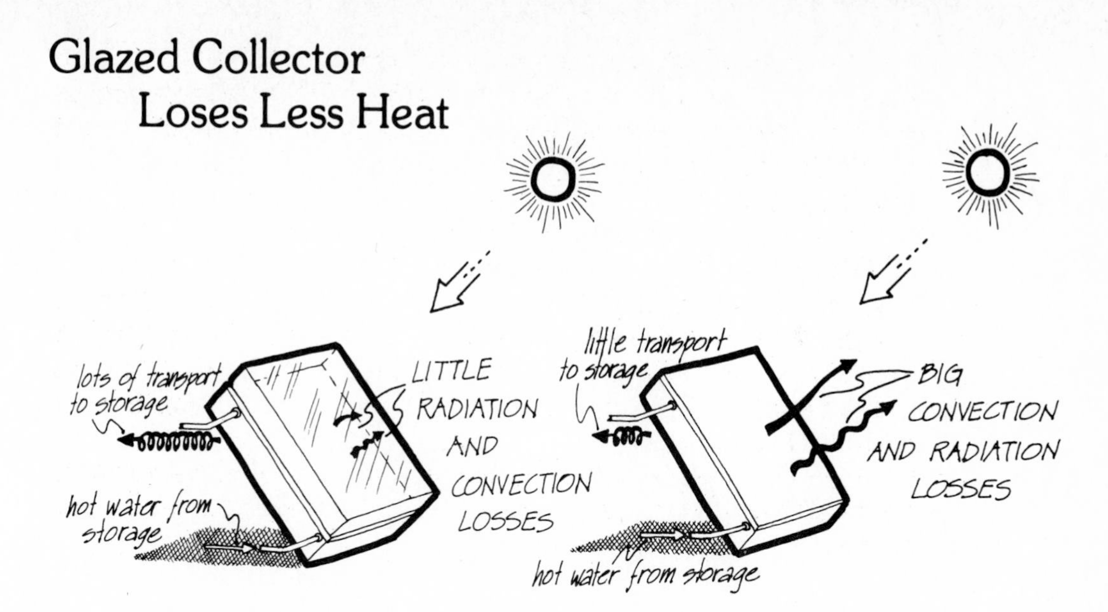

Whether or not glazing is needed depends simply on the temperature required from the solar heater and the outside temperature at which it must operate. Solar house heaters and solar hot-water heaters must deliver high temperatures and usually operate in cold weather, and they are commonly glazed. Solar pool heaters don't need to deliver hot water and usually operate in mild weather, so they are commonly unglazed. However, pool heaters used in winter are often glazed, and hot-water heaters used in warm climates (parts of Israel, for instance) are often left unglazed.

NIGHTTIME HEAT LOSS 15

The solar collector has captured the sun's heat during the day, and the captured heat has been pumped to storage. But what happens when the sun goes down or the sky clouds over? Again, we'll use our rainwater apparatus to understand what happens.

Suppose the rainwater tank is nearly filled and the rain stops. What will happen? With no rain falling on the rain collecting tray, no rainwater flows down the pipe. But water can still flow into the pipe *from the tank* through the tube that connects the two. The water depth in the pipe, then, will have almost the same depth as the water in the tank, since water can flow easily between them. However, when there is water in the pipe it leaks out—the higher the depth, the more the leakage.

Tank Loses Water when Rain Stops

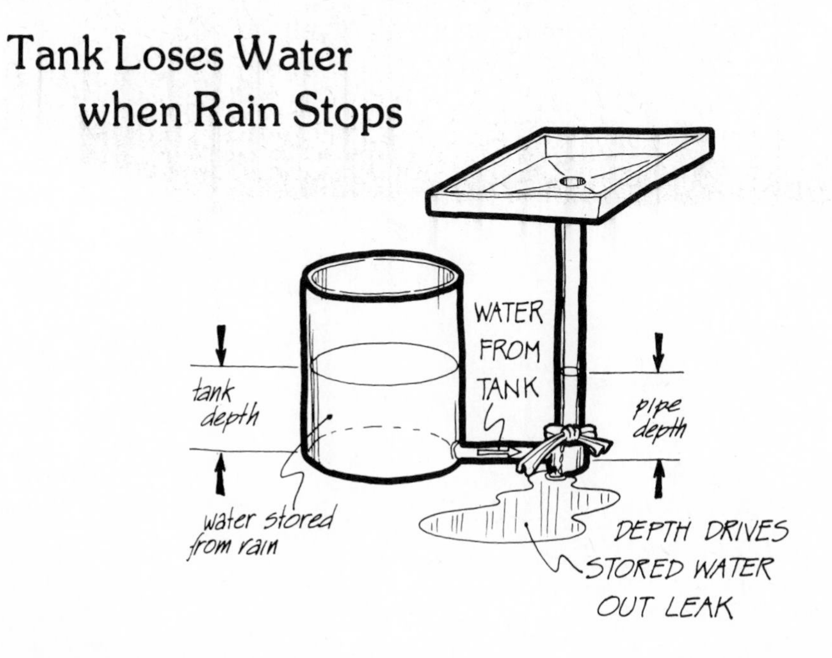

The net result is that stored water can flow *out* of storage and *into* the pipe and then flow out the leak. All the carefully collected rainwater leaks away as soon as the rain stops.

To prevent loss from the tank when the rain stops, we can put a *check valve* in the tube connecting the pipe and the tank. A check valve is a device that blocks the flow of water in one direction but allows it to flow in the other direction. Check valves are common devices to let a liquid flow in only one direction; in fact, your own heart has several such valves to keep your blood flowing in the right direction.

Water Flows One Way through Check Valve

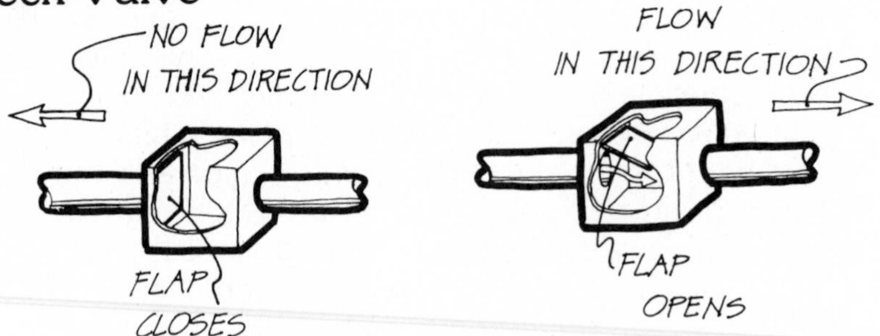

With a check valve inserted in the tube between the pipe and the tank, rainwater can flow easily from the pipe to the tank when the tray is catching rainwater. But when the rain stops, the check valve prevents any stored water from flowing back out the tank and into the pipe where it leaks away.

Check Valve Prevents Leakage when Rain Stops

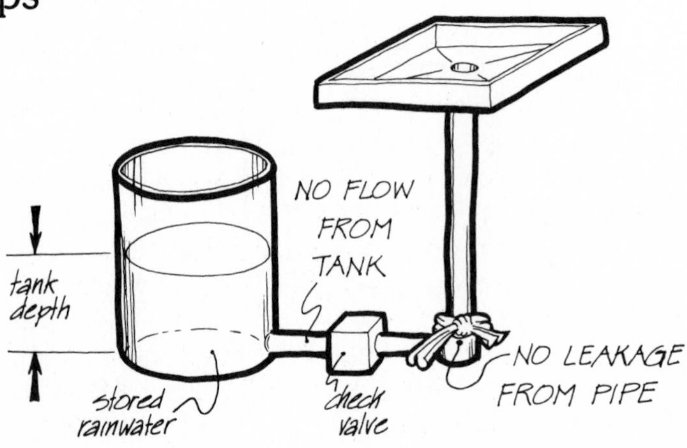

The same process can occur in a solar heating system. Heat in the heat storage tank can be lost after the sun goes down. When solar radiation is not being absorbed by the collector plate, the water *loses* heat as it passes through the collector; it does not gain heat. Hot water pumped from the storage tank to the collector is hotter than the outdoors temperature, so heat flows by radiation and convection to the outdoors.

Storage Loses Heat at Night

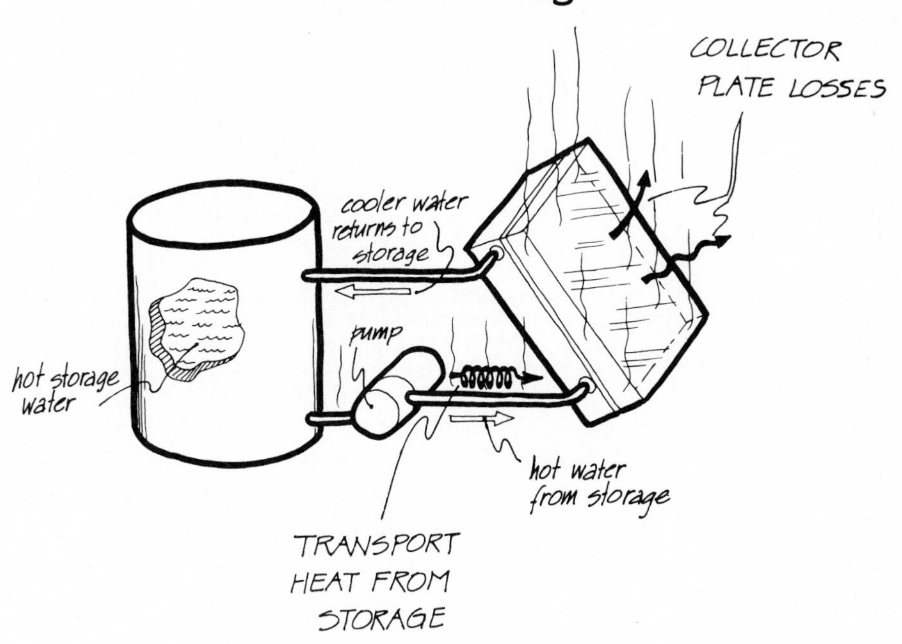

As stored water from the rainwater collector tank leaks away when the rain stops, so stored heat leaks from the solar collector tank when the sun is not out. In the rainwater system, we can use a check valve to prevent rainwater from flowing back from the tank to the pipe. The check valve lets rainwater *into* the tank, not out of it where it can leak away. In our solar heating system, the same effect can be achieved simply by stopping the pump whenever the sun isn't shining. Just as the check valve senses when the tank depth is greater than the pipe depth—in other words, when rainwater is leaving the tank—temperature sensors detect when the heat storage tank is hotter than the collector—when heat is leaving storage—and they control the pump. If the storage tank is hotter than the collector, the pump is shut off; if the collector is hotter than the storage tank, the pump is turned on.

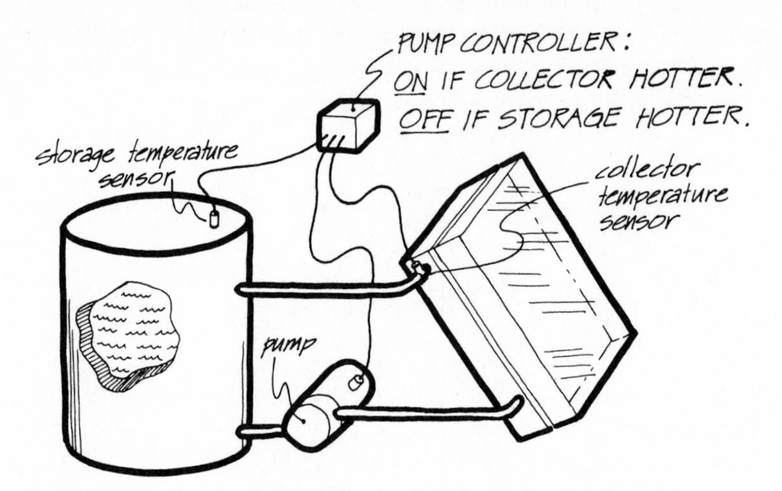

Controller Prevents Nighttime Heat Loss

A TYPICAL DAY: MORNING 16

Now that we've examined all the elements of a solar heating system, let's see how it operates during a typical winter day. Right now we'll consider only collecting and storing heat; later we'll see how we use the stored heat. We'll also follow the operation of our rainwater collecting apparatus so comparisons can be made between the two at various stages.

Let's assume that both collecting systems begin the day partly full. For example, in the rainwater collector, there might have been some rainwater left from previous rain. In the solar collecting system, we'll assume the heat storage tank is still warm from the previous day's collection.

In the rainwater apparatus, the check valve has closed to keep the stored water from leaking out. No rain is falling, and any water in the pipe has long since leaked out. As for the solar heater, the sun has not yet risen. The collector, which has been exposed to the cold night air, is much colder than the heat storage tank. The heat storage is warmer than the outdoors because it still contains heat. The controller had sensed that the collector was colder than the heat storage tank and shut off the pump, preventing the storage tank from cooling.

As the sun rises, the collector starts to get a little warmer. But since a south-oriented collector doesn't face the sun early in the morning, not much solar radiation is absorbed. The collector will get warmer, but not as warm as the heat storage. Since the collector is still colder than the storage tank, the controller won't turn on the pump. Of course, the collector is a little warmer than the outdoors, so it loses heat through its glazing. In fact, it loses *all* the heat absorbed by the collector: since none is saved, all of the incoming heat is lost.

For the rainwater apparatus to be analogous to our solar collector, let's suppose that it starts to drizzle on the rainwater

Before the Sun Rises

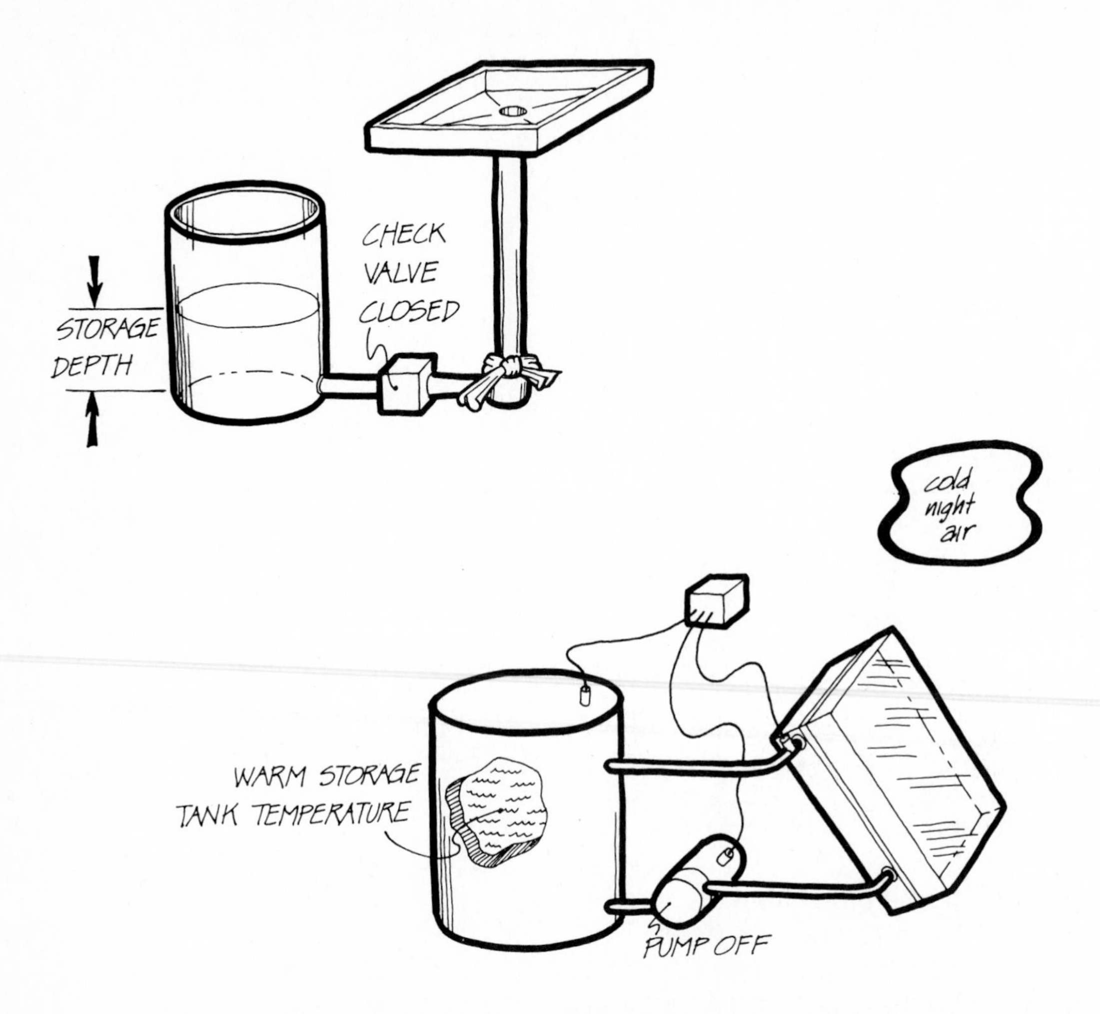

collecting tray. Since little rainwater is being collected, the water in the collection pipe dosn't get as deep as the storage depth. No rainwater flows into storage, and efficiency is zero—all the rain captured by the tray leaks away.

Until the temperature of a collector gets hotter than the storage tank, no heat is collected. Often on cloudy days, a solar collector doesn't collect any heat whatsoever, just as when the sun first rises in the morning not enough solar radiation falls on the collector to make its stagnation temperature hotter than the storage tank's. The heat leaks away from the collector as fast as it is absorbed. The cold temperature outdoors aggravates the situation, since the incoming solar radiation must further raise the collector's temperature so that it equals the storage temperature.

Just After Sunrise

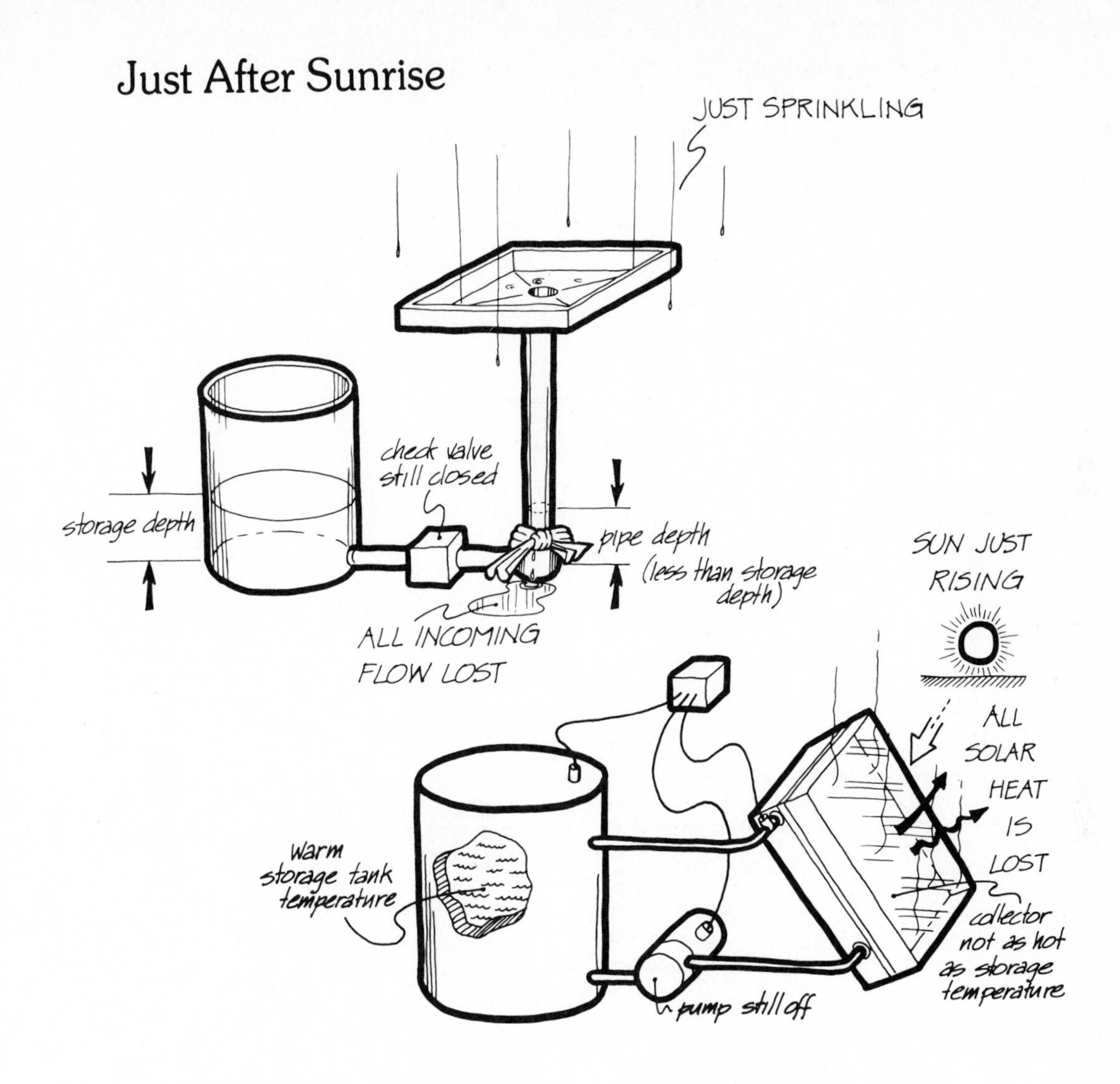

As the sun gets higher in the sky, its radiation strikes the collector more and more head-on, and more solar radiation is absorbed by the collector plate. Eventually the collector plate gets hotter than the storage tank. When it does get hotter, the pump controller senses the temperature difference between the collector and the storage tank and turns on the pump. With pump water flowing, some of the heat absorbed by the collector is moved into storage: water flowing from storage is heated as it passes through the collector. Warm water from the collector further heats up the storage tank. Now not all the collector's heat is lost, since some of it is going into storage.

In the rainwater collection system, let's suppose the rain increases from a drizzle to a light rain; this is analogous to more sunshine being absorbed by the solar collector. With more rainwater collected and flowing down the pipe, the pipe's water depth gets higher than the storage water depth. The check valve opens and not all the rainwater is lost through the leak, since some is flowing into storage.

Early Morning

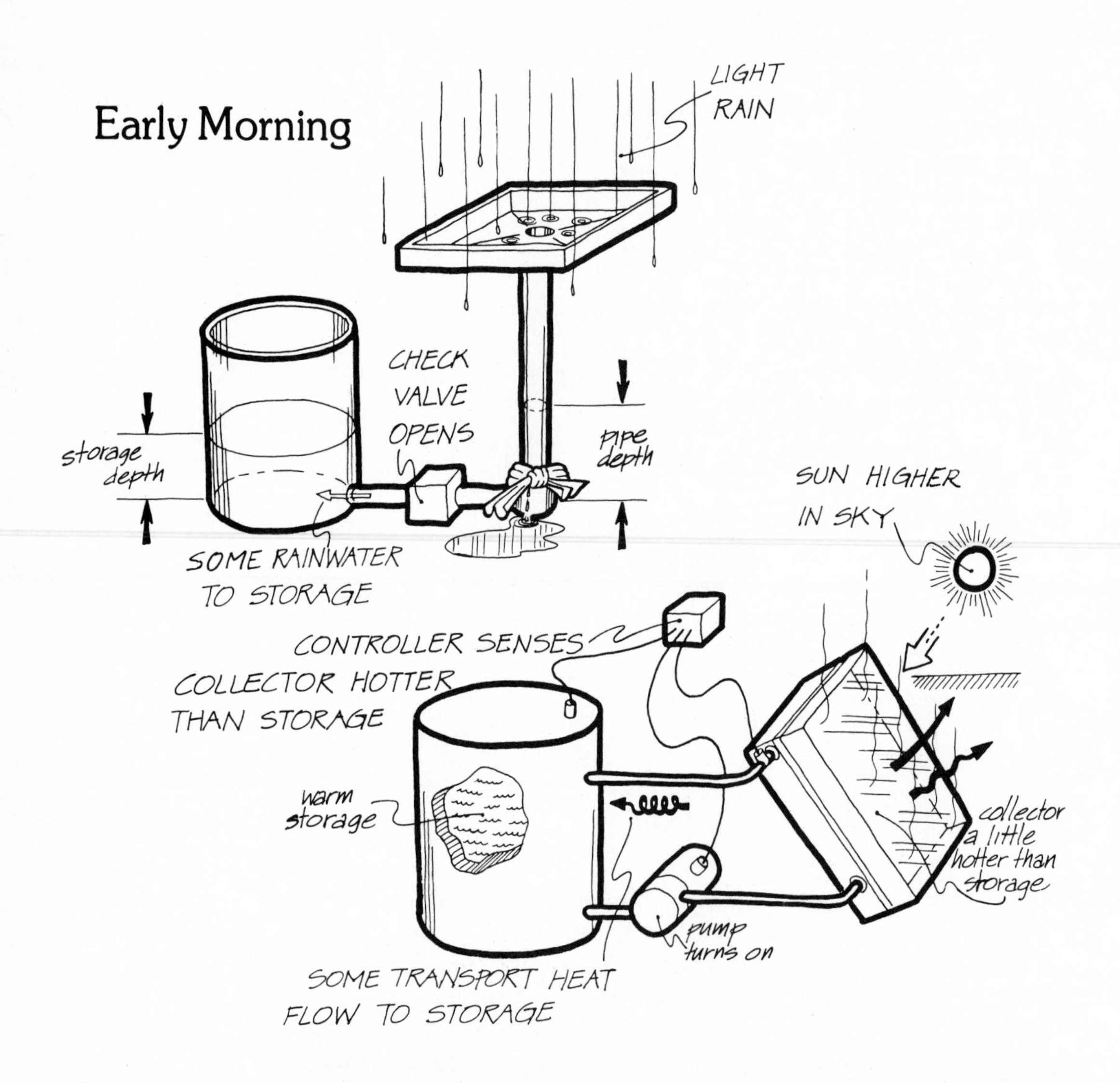

In the late morning the sun faces the collector nearly head-on; and much more solar radiation is absorbed by the collector plate. With the pump on, warm water from storage is pumped through the collector, where it is further heated. It returns to the storage tank hotter and continues to heat the tank. The collector always stays a little hotter than the storage tank, so the controller always keeps the pump on. Efficiency is high in the late morning. The collector receives water from storage, and before noon storage isn't very hot yet. When the collector isn't hot, heat losses are low, resulting in high efficiency.

The rainwater equivalent to the sun hitting the collector nearly head-on is a heavy rain. The incoming water fills up the pipe and flows through the check valve into the storage tank. Since the pipe's water depth isn't too high yet, not much rainwater is lost and efficiency is high.

Late Morning

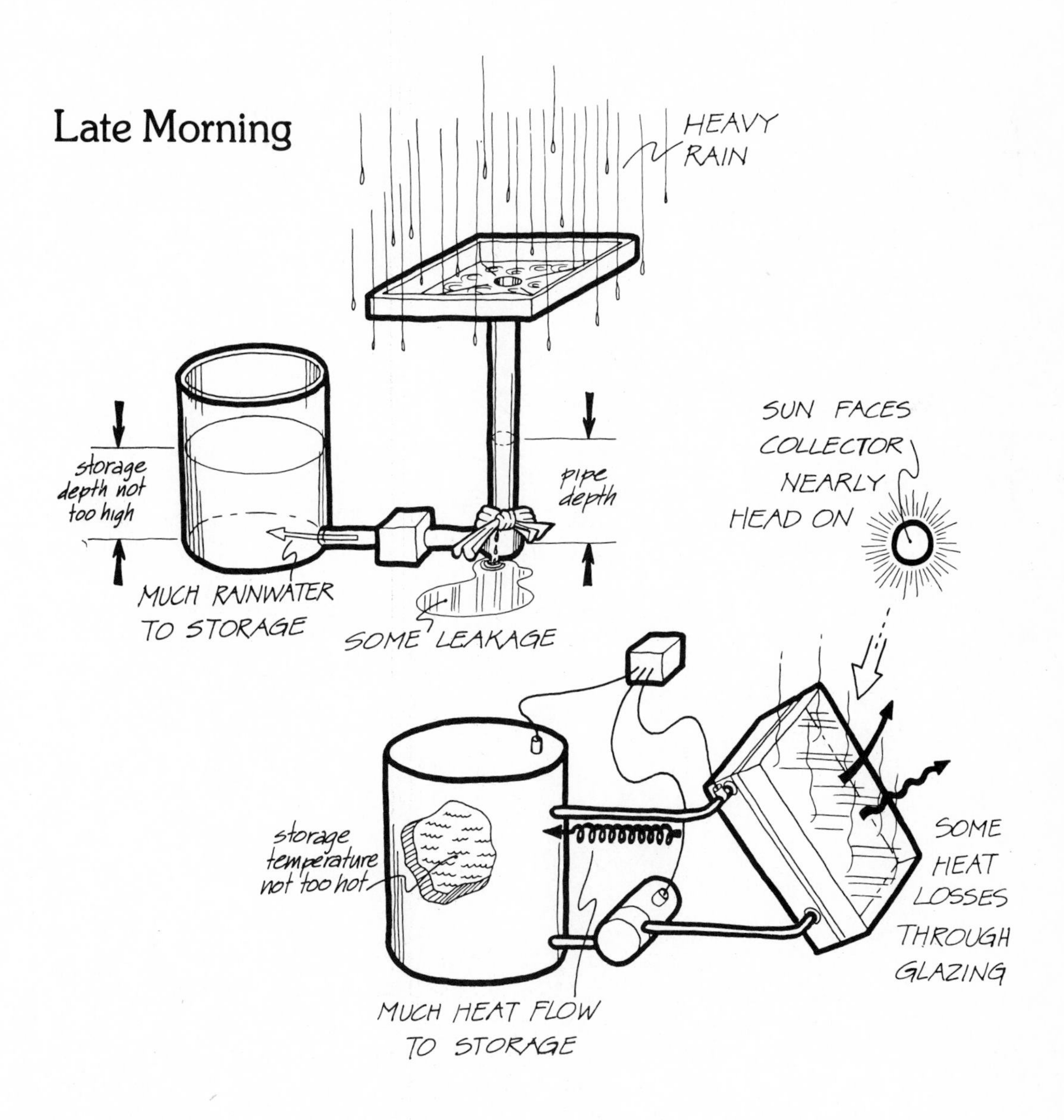

A TYPICAL DAY: AFTERNOON 17

The midday hours (10:00 A.M. through 2:00 P.M.) are the most important for solar collecting. During these hours the sun faces the collector almost head-on. A large fraction of the solar heat collected in a day occurs during these midday hours. By contrast, the early morning sun doesn't face the collector squarely, and the same is true in the late afternoon. In addition, the early morning and late afternoon sunlight has to pass through more of the atmosphere on its way to the collector.

More Solar Radiation Near Mid-Day

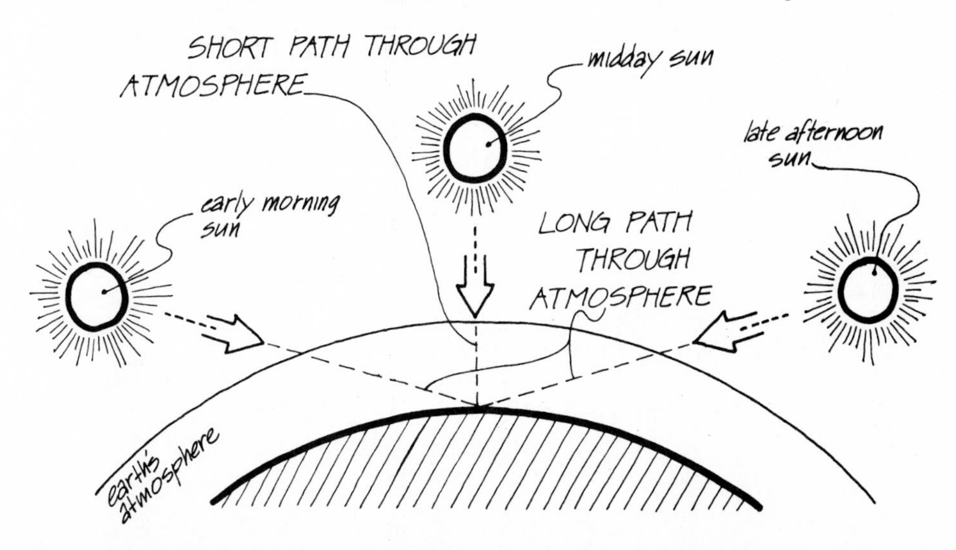

Solar radiation is diminished as it goes through the atmosphere, so less of it reaches the collector during the early morning and late afternoon. During the midday hours the sun angle is better, and the sunlight passes through less atmosphere as well. In fact, in winter, a flat-plate collector can collect almost as much heat as a *tracking* collector, which has a tracking mechanism to keep the collector always pointed at the sun. Even though the tracking collector can face the early morning and the late afternoon sun, there isn't much solar radiation to be collected then anyway. One that doesn't move at all can collect almost as much heat, yet be simpler and cost less.

A few hours past noon, most of the solar heat to be stored during the day has been moved by transport heat flow to the storage tank. The water in storage won't get much hotter, since little more heat will be added. Enough solar radiation is still present to heat the collector so that it is slightly hotter than the storage tank. But efficiency has fallen off: the collector is much hotter than the outdoors, making losses high. High losses mean little heat collected and lower efficiency.

In the rainwater system, a heavy rain is still falling—analogous to the sun being nearly square to the solar collector. But after heavy rain for a long time, the storage water depth and the water depth of the pipe are much higher. The pipe's water depth

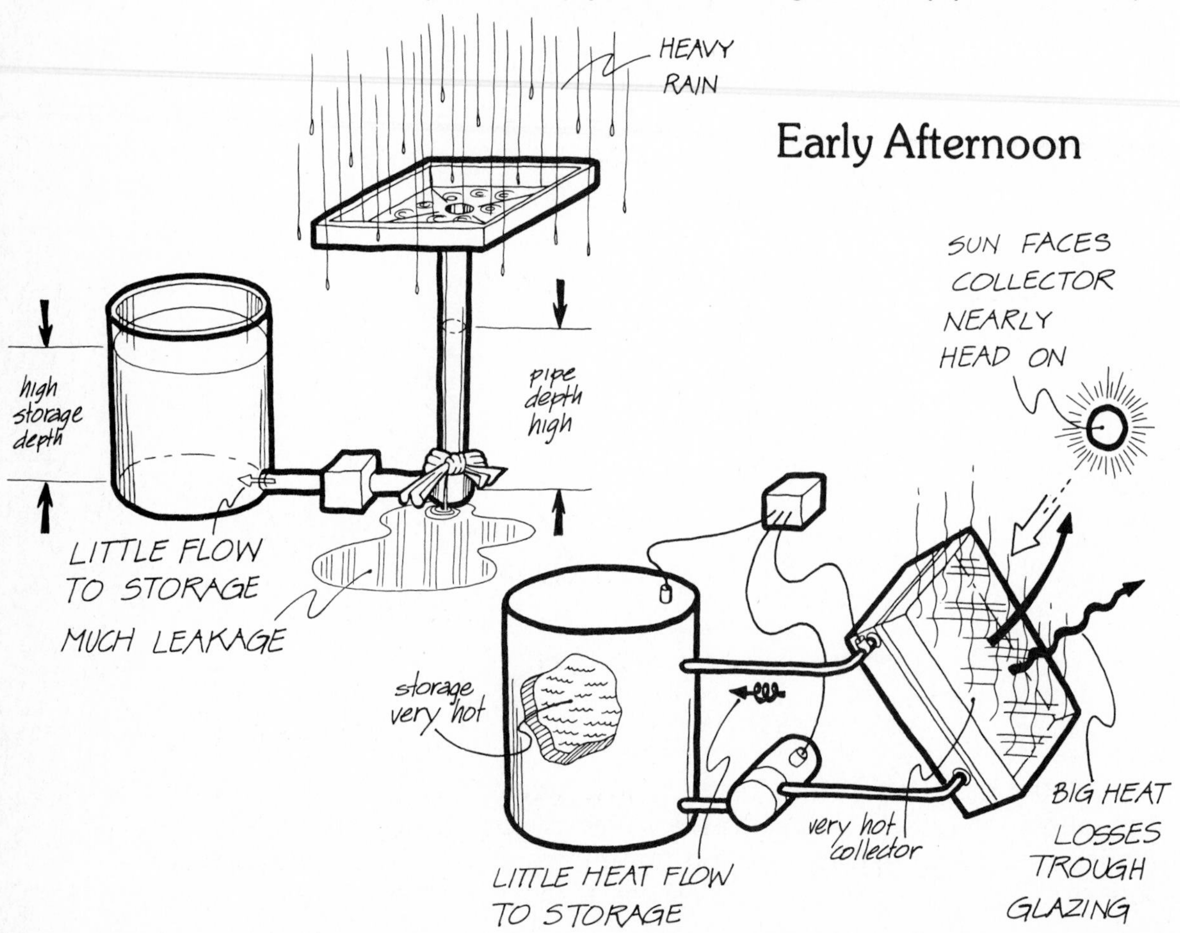

is high enough to force much of the rainwater out of the leak. High leakage leaves little of the incoming flow to go to storage, so efficiency is lower.

Late in the afternoon the sun faces the collector less squarely, and less solar radiation is transmitted through the atmosphere, as we discussed earlier. The net effect of these factors is that the solar radiation absorbed by the collector falls off rapidly. But losses are still high, since they are determined by how hot the water coming from storage is. Eventually the collector can't absorb enough heat to stay hotter than the storage tank. When the collector gets cooler than the storage tank, the controller shuts off the pump. Of course, efficiency is zero when the pump is off. Any incoming solar radiation that is absorbed is lost completely through the glazing.

Let's suppose that the heavy rain falling on our rainwater tray diminishes, just as the solar radiation absorbed by the solar collector in the late afternoon is reduced. Less rainwater flows down the pipe, but there's still a lot of leakage since the pipe's water depth is so high. The reduced inflow can't maintain as

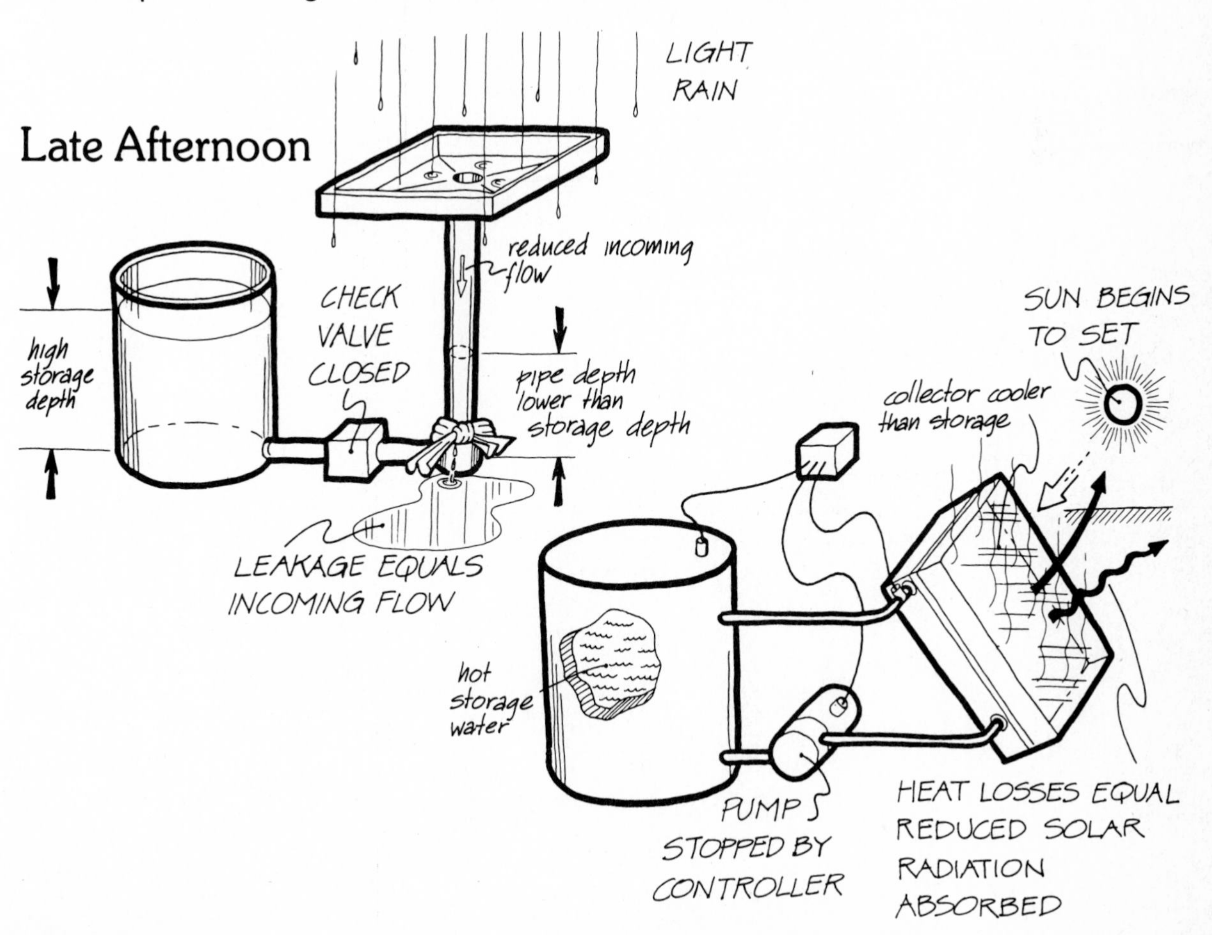

high a stagnation depth in the pipe because so much is leaking out. When the pipe's water depth drops below the storage water depth the check valve closes, preventing loss of our carefully stored rainwater out through the pipe's leak. As soon as the check valve closes, efficiency drops to zero: all the rainwater flowing down the pipe leaks away.

Finally the sun sets. The solar collecting system appears from the outside just as it did before the sun rose that same morning. The collector, exposed to the cold night air, is much colder than the heat storage tank. No water is pumped from storage to collector because the controller has long since sensed that the storage is hotter. Only the temperature of the storage water has changed. Whereas before dawn it was only warm, after sunset it is much hotter.

Similarly, when the rain stops falling on the rainwater collector, the only difference is the amount of stored rainwater. The water in the pipe has long since leaked out and the check valve is closed, preventing loss of the stored rainwater. The net effect of the rain falling has been simply to store some portion of the rain caught by the collecting tray. Although some rainwater leaked away, the storage water depth after the rainstorm is much higher than it was before.

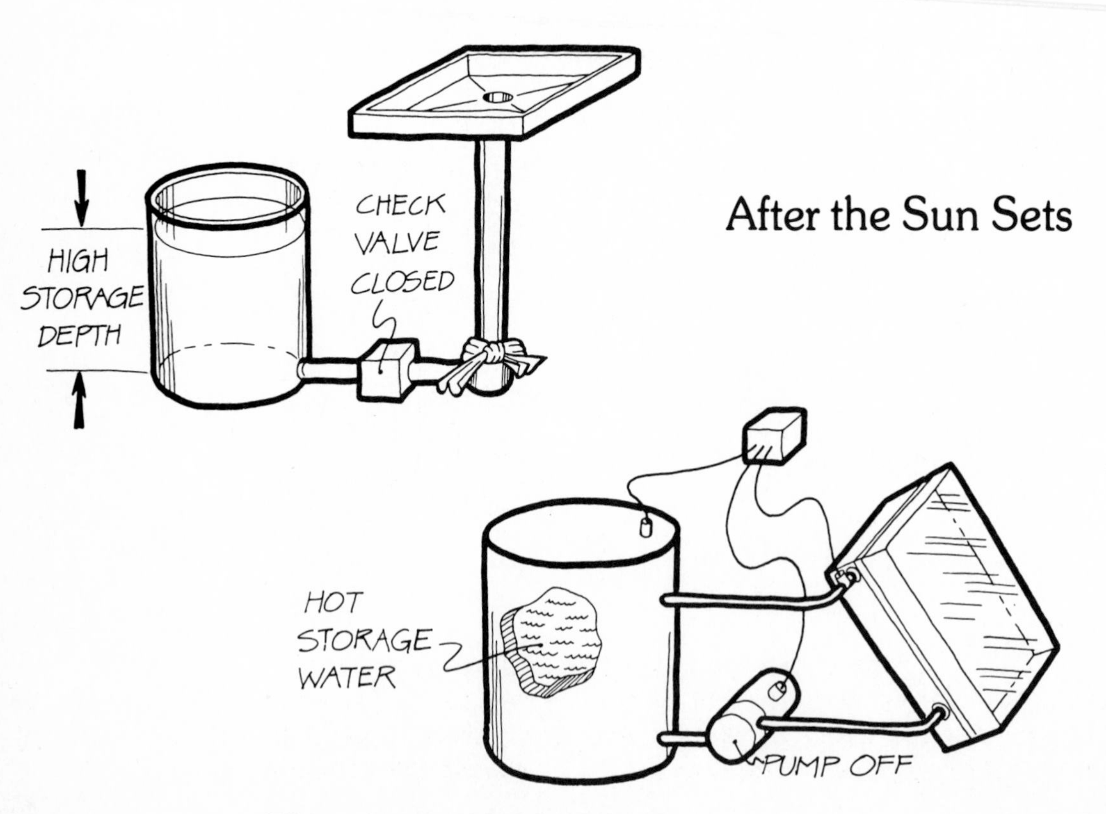

USING STORED HEAT 18

Solar heat has been collected and stored in order to be used. In a swimming-pool heater it's easy to use the stored heat. Instead of pumping the solar-heated water to a heat storage tank, the pool itself stores the sun's heat. But if we're trying to heat a house or water, we need some way to get the heat out of storage. By analogy, if we want to get the rainwater from the rainwater collector, one way would be to tap into the side of the storage tank. We can tap off some stored rainwater whether we're simultaneously collecting rainwater or whether the rain has stopped and we're only tapping off the stored rainwater.

Using Stored Rainwater

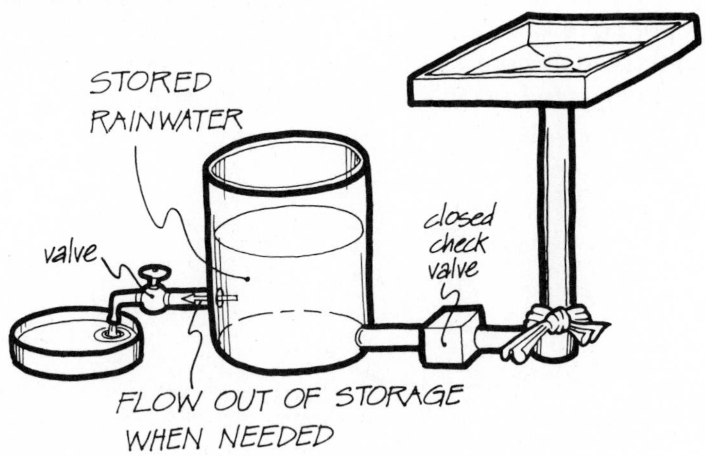

Similarly, a heat storage tank can be "tapped" to yield its stored heat. However, instead of actually removing the hot water from the storage tank, a *heat exchanger* is usually used to extract the heat—but not hot water. A heat exchanger consists of a coil of copper tubing which sits inside the heat storage tank; cool water is pumped into one end of the coil and hot water flows out the other end. The stored hot water heats the outer surface of the copper coil by natural convection; the heat flows easily through the copper by conduction. The water flowing through the coil is then heated by convection from the hot copper surfaces that the water flows past. This kind of heat exchanger is called a *liquid-to-liquid* heat exchanger, since heat is transferred from one liquid (the hot storage water) to another (the water pumped through the coil).

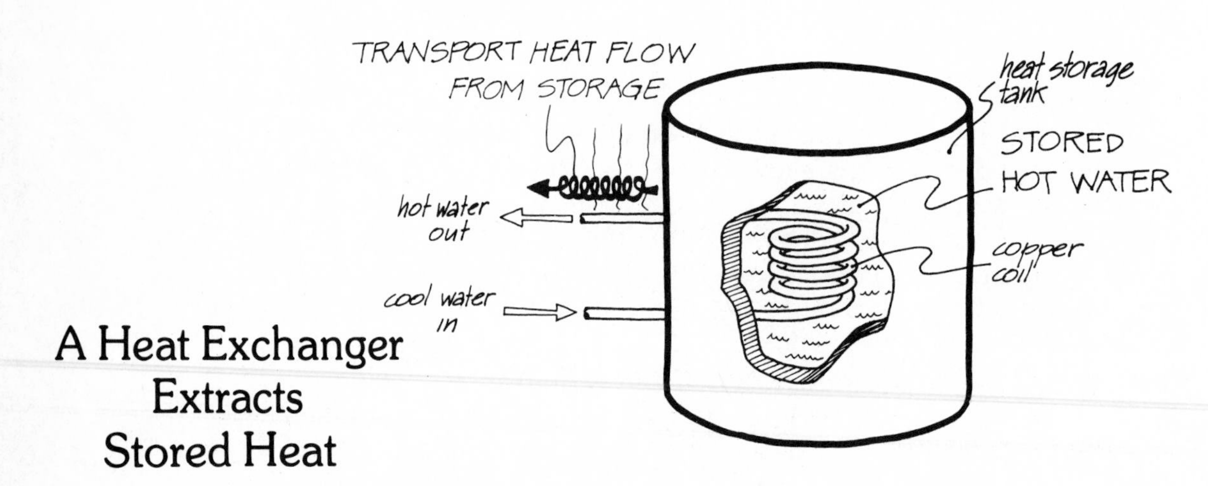

A Heat Exchanger Extracts Stored Heat

Cool water can be pumped into the heat exchanger whether or not the solar energy system is also collecting heat. It is possible to withdraw heat from storage while simultaneously adding solar heat to storage. The only requirement for the heat exchanger is that the stored water must be hotter than the cool incoming water.

Using Stored Heat

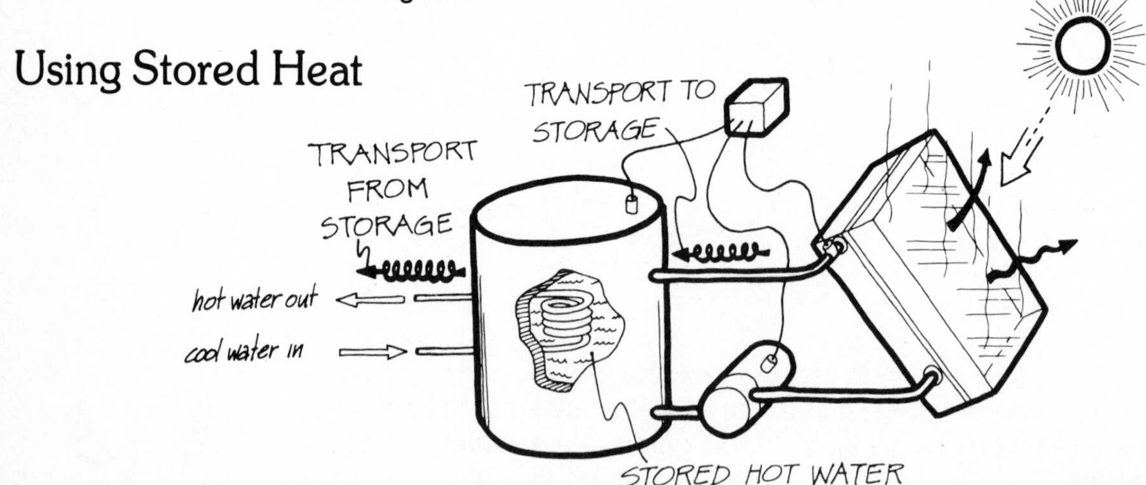

Let's consider the case of heating water. Usually a solar heating system is used as a *preheater* for a conventional gas, oil, or electrical hot-water heater. Cold water from a well or water main flows through the heat exchanger before going to the conventional *auxiliary* unit, as it is sometimes called.

Solar Water Heating: Preheater Arrangement

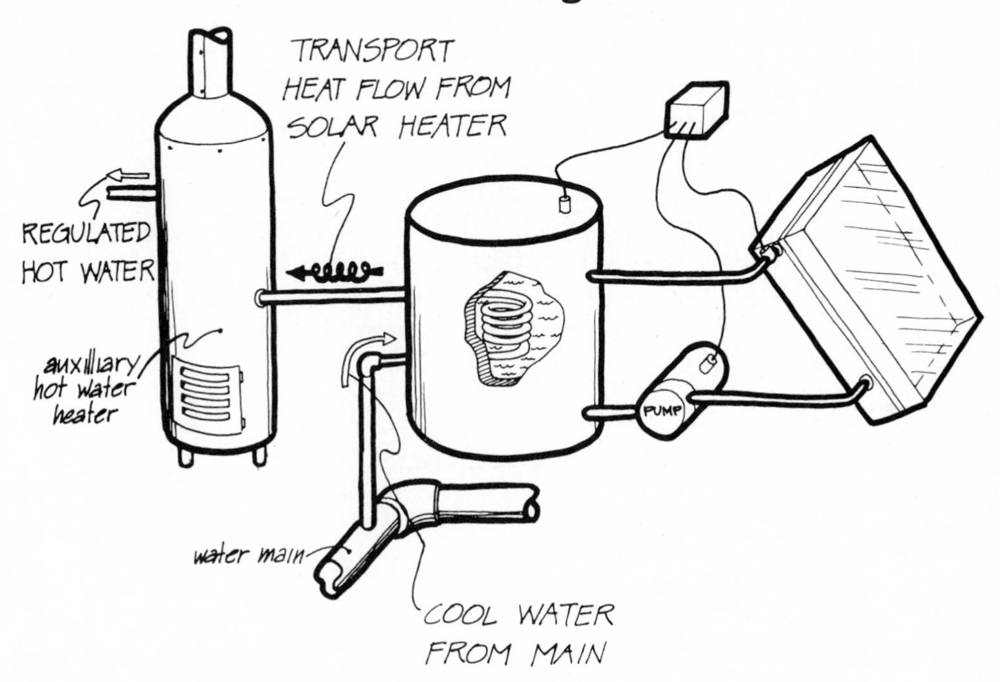

The preheater arrangement has several advantages. First, the stored solar-heated water only has to be hotter than the water main temperature to extract stored heat. Usually, the water main temperature is about the same as the ground temperature, so even a little sunlight can heat the water from the main. Second, if the solar heater cannot get the water hot enough, the auxiliary hot-water heater can *top it off*. In the winter, for example, the water main temperature could be 50° F. The solar heater might only be able to heat it to 100° F. The auxiliary heater needs only to heat it another 30° for its temperature to reach the desired 130° F. Most of the heat was added by the solar heater, but some was added by the auxiliary heater. Since most people are accustomed to regulated hot water, having a conventional auxiliary unit in addition to the solar preheater is almost a necessity.

Similarly, in house heating, a solar heating system can be used as a preheater for a conventional furnace. Let's suppose we're solar heating a house that is heated by *forced air* (forcing heated air through air ducts). We can extract solar heat from our storage tank with a *liquid-to-air* heat exchanger.

Recall that we've already used a liquid-to-*liquid* heat exchanger to extract heat from our storage for hot-water heating. The two types of heat exchangers look distinctly different. This is because it is so much easier to transfer heat to water than to air, and much more surface must be exposed to the air than to the water.

In a liquid-to-air heat exchanger (a car radiator is one type), hot water pumped through many parallel tubes transfers heat to the air by way of hundreds of thin metal fins attached to the tubes. Heat flows easily into the metal tubes from the flowing hot water. Then it's conducted easily along the fins, insuring that lots of hot fin surface is exposed to the air blowing by. Because heat doesn't flow easily by convection from a surface into air, a lot of fin surface is needed and the air must blow fast over the fins.

A Liquid-to-Air Heat Exchanger

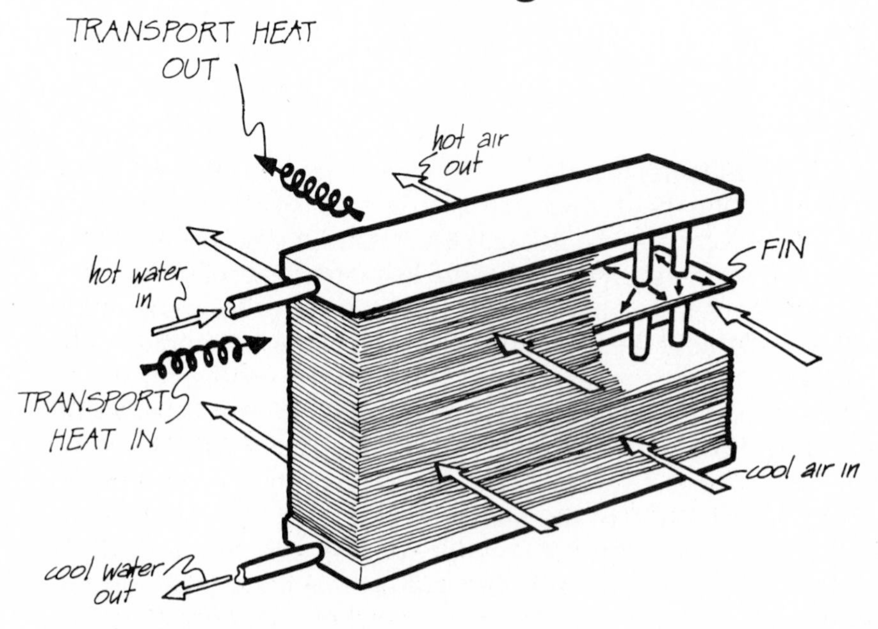

In heating a house with a solar preheater, a liquid-to-liquid heat exchanger draws heat from the solar heat storage and feeds it into a liquid-to-air heat exchanger. The latter transfers heat to the air flowing into a furnace—often an electric furnace. If the air is sufficiently hot, the furnace's coils never come on. But if the solar heat storage is too cool, the coil of the furnace will add additional heat to the air.

Solar House Heating: Preheater Arrangement

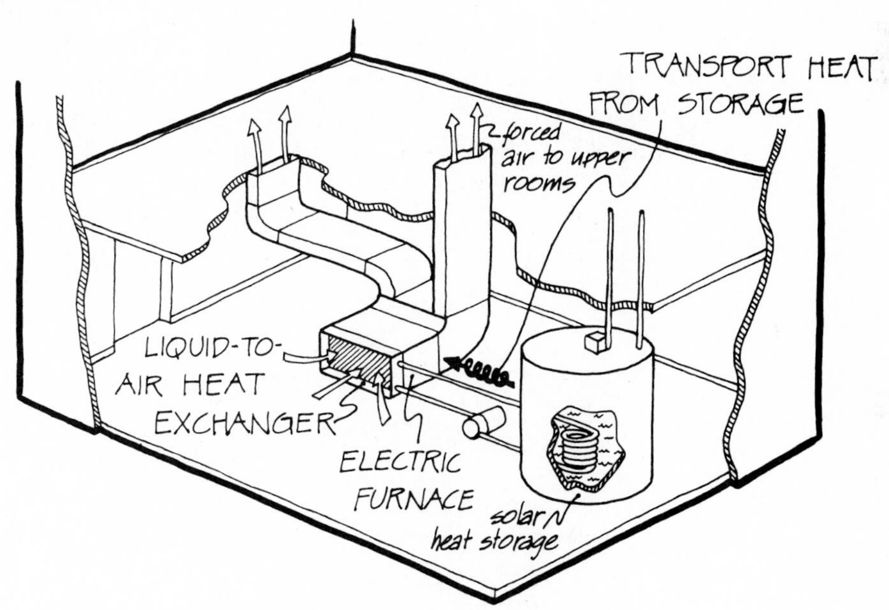

Solar heat is also used as a preheater for a conventional heating system. Note that even though electricity is a very expensive fuel, it is used only a fraction of the time: the house is primarily solar heated, and electricity is used as a back-up.

So far we've talked about preheater, or *series* arrangements of solar and auxiliary heating systems. Series, in electrical terms, means flowing first into one and then into the other; the auxiliary adds its heat only after the solar heat is added. Another arrangement is a *parallel* one where both the solar heating system and the auxiliary unit directly heat the house. An example of a parallel arrangement is a house that is simultaneously heated by electric baseboard heating and hot-water baseboard heating. Both systems add heat to the house, but the electric auxiliary system comes on only when the temperature of the solar storage is insufficient to heat the house.

Solar House Heating: Parallel Arrangement

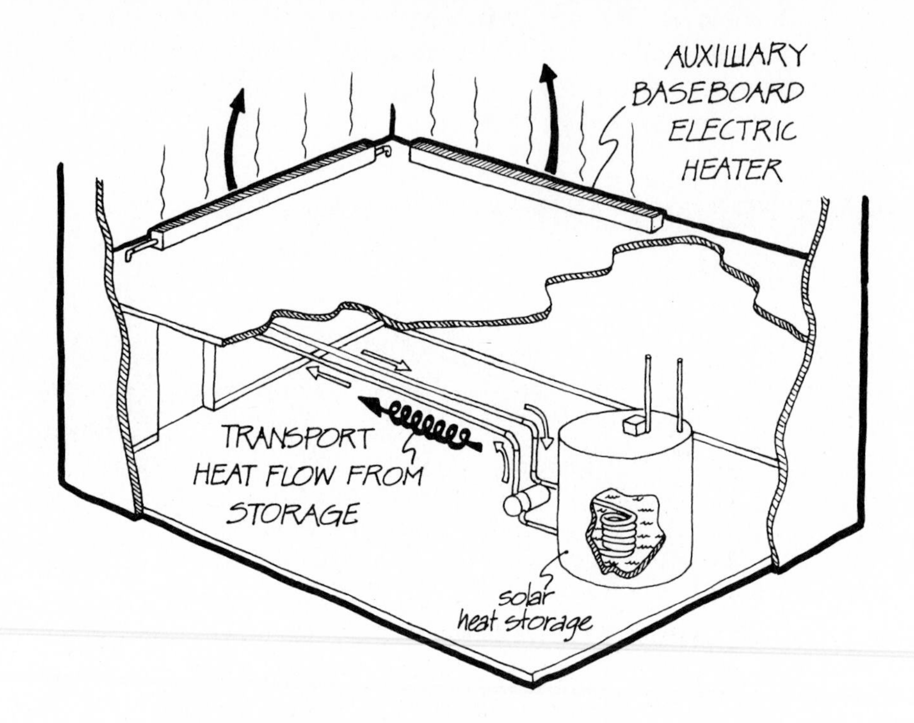

There are many other ways that solar energy can be coupled with an auxiliary unit for house and hot-water heating. The ones shown here are just a few of the possibilities.

But why have an auxiliary unit at all? Without one you couldn't count on having your hot water or house heated after several days of cloudy weather. Then why not store a week's supply of heat? From an economic viewpoint, storing much more than a day or two's heat supply becomes expensive. A storage tank that holds a week's supply of heat costs much more than one that holds a day's supply. Although some solar energy heating systems have been built to store heat for months at a time, the majority of residential systems store only a day's heat supply.

AIR-HEATING SOLAR SYSTEMS 19

So far we've only discussed *hydronic*, or *water-heating*, solar energy systems. It's also practical to let solar radiation heat *air* instead of water. An *air-heating* solar collector is very similar to the water-heating collectors we've already learned about. Air, instead of water, is forced past solar-heated surfaces.

Most air-heating solar collectors are made with an absorber surface with air ducts instead of water ducts attached to the back side. As in a water-heating collector, glazing prevents heat loss from the black absorbing surface. An inlet duct lets air be blown into one side of the collector, and an outlet duct lets it flow out the other side. Insulation prevents heat loss from the sides and back of the air duct.

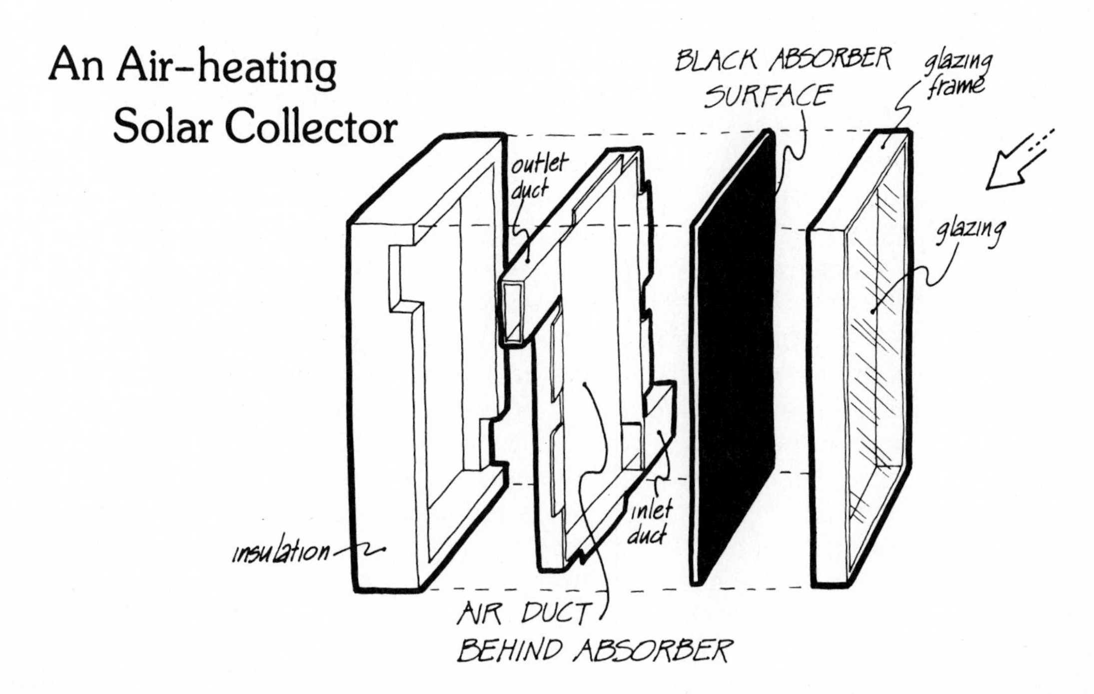

After solar radiation passes through the glazing, it is absorbed by the black absorbing surface. Some heat is lost by natural convection and radiation from the absorbing surface, depending on how hot it gets—as in a water-heating solar collector. What is not lost flows by conduction through the absorbing surface to the air duct where it flows by forced convection into the airstream and is transported to storage.

Heat Flow in an Air-Heating Collector

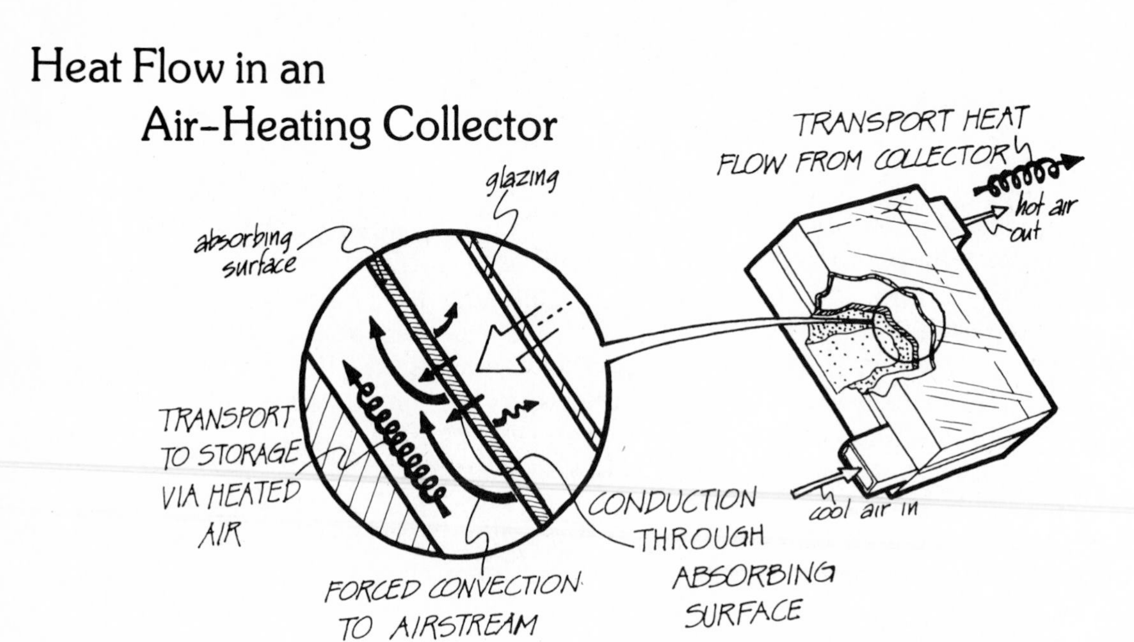

In air-heating solar collectors, lots of surface area must be exposed to the airstream. Often the back side of the absorbing surface has many fins, as do liquid-to-air heat exchangers and baseboard heaters. Heat conducted through the metal absorbing surface is conducted along the fins. With lots of fin surface exposed to the air in the ducts, solar heat can easily flow by convection into the airstream.

Storage for an air-heating system is a *rockbed*—a closed container of crushed rocks, each about the size of an egg. As the hot air flows through the crevices in the rockbed, it gives up its heat to the rocks. The rocks heat up and the air cools. If much smaller rocks are used, it's hard for the air to flow through the crevices; if much larger rocks are used, they don't heat up thoroughly. Heat in the rockbed *stratifies*; the rocks higher up are hotter than those near the bottom of the bed. Air coming from the bottom of the rockbed is cooler than the solar-heated air entering the top.

Fins Improve Air-Heating Collectors

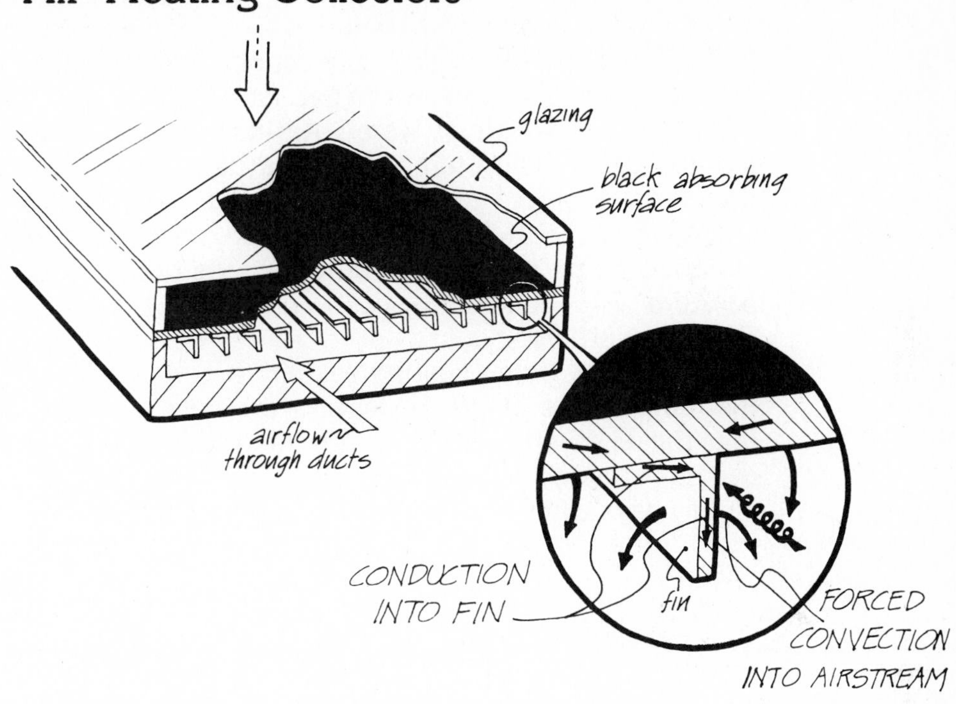

Storing Heat in a Rock-Bed

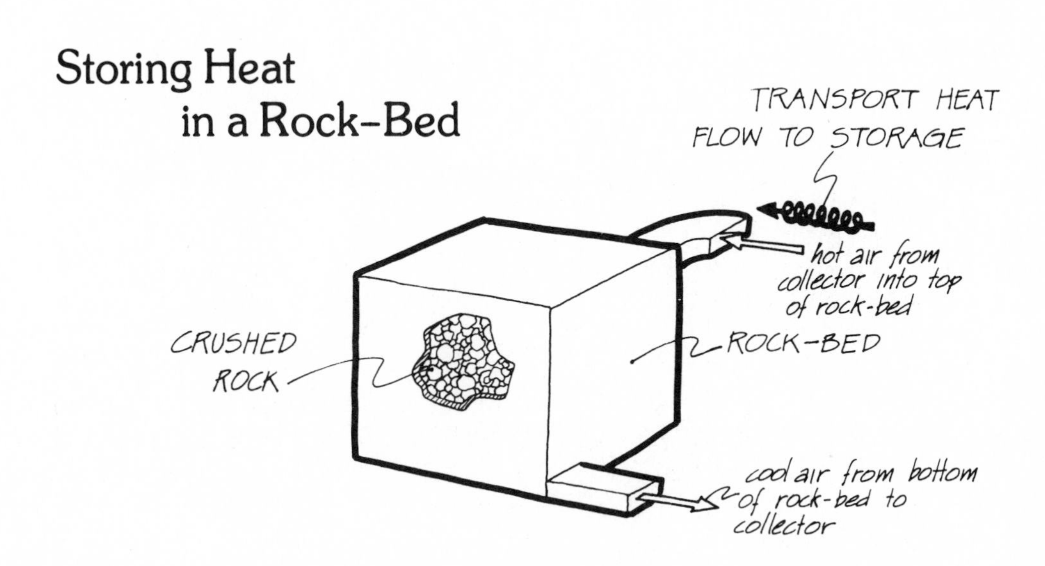

Functionally, the air-heating system works almost exactly the same as a water-heating system does. Solar heat is transmitted through the glazing and is absorbed by the black absorbing surface. The heat is conducted from the absorber surface and convected by air flowing through air ducts. Similarly, in a hydronic system, heat is also conducted from the absorber surface but is convected by water pumped through tubes. In both types of collector, heat is lost from the absorbing surface by means of radiation and convection, depending on how hot the surface gets. The heat that isn't lost is transported to storage by either air or water. As in a water-heating system, storage increases in temperature as heat is added. Cooler air (rather than cooler water) is blown (rather than pumped) back to the collector. Efficiency—the fraction of the incoming solar radiation going to storage—still depends on the collector temperature: a hot collector gives low efficiency and a cool collector gives high efficiency.

As in a water-heating system, a controller detects when the temperature of the air leaving the solar collector is hotter than the air entering it. The controller turns on the fan when heat can be extracted from the sun, and it turns it off at night or on cloudy days when no useful solar heat is available.

An Air-Heating Solar Energy System

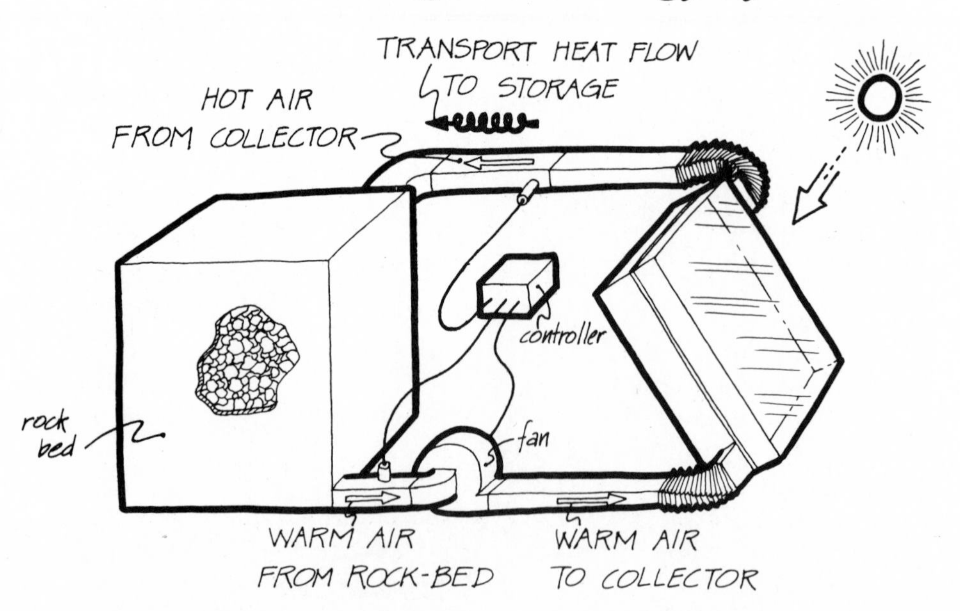

All in all, air-heating systems function much the way water-heating systems do. In fact, the systems are so similar that the rainwater analogy can represent both. However, although they are functionally similar, there are important differences in the design of the two types of systems.

One way solar air-heating collectors differ from water-heating ones is in *temperature rise*. Temperature rise is the difference between how hot the air is going into the collector and how hot it is coming out. Since air doesn't transport heat as well as water does, air flowing through a collector tends to heat up more. If the same amount of air and water were flowing into their respective collectors at the same temperature, the air would come out hotter than the water. If the air comes out hotter, then the *average* collector temperature would be hotter too—and a hot collector means low efficiency.

Fortunately, air-heating systems *don't* have low efficiency, since the air blowing into them is cooler. Because of the stratified effect mentioned earlier, air flowing into the collector comes from the rockbed fairly cool. Cool air entering the solar collector tends to lower the collector's temperature and to improve efficiency. The improvement due to the cool inlet of air from the rockbed offsets the reduced efficiency of a big temperature rise through the collector: the air starts off cooler but gets hotter. The net effect is that an air-heating system is just as efficient as a water-heating system.

What are the advantages and disadvantages of using an air-heating system? Generally, air systems are best for house heating, while hydronic systems can be used for house heating, hot-water heating, and pool heating. Although it is possible to make an air-heating hot-water heater or pool heater (using an air-to-liquid heat exchanger), it is usually too costly to be practical.

For house heating, an air-heating system won't spring a leak and ruin your rug as a water-heating system might do. Leaks in hydronic systems can be caused by corrosion of metal

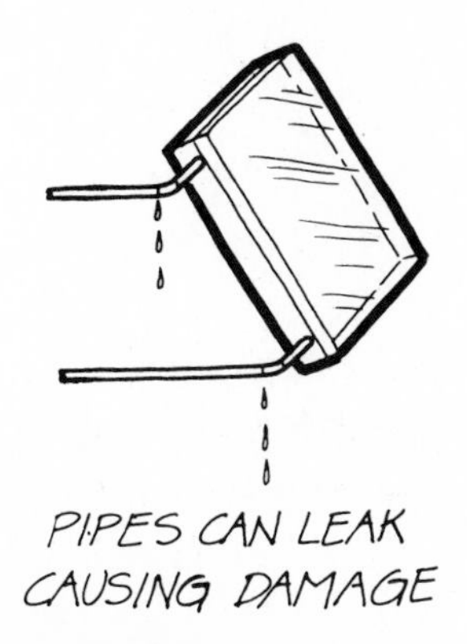

Water-Heating and Air-Heating Systems can Leak

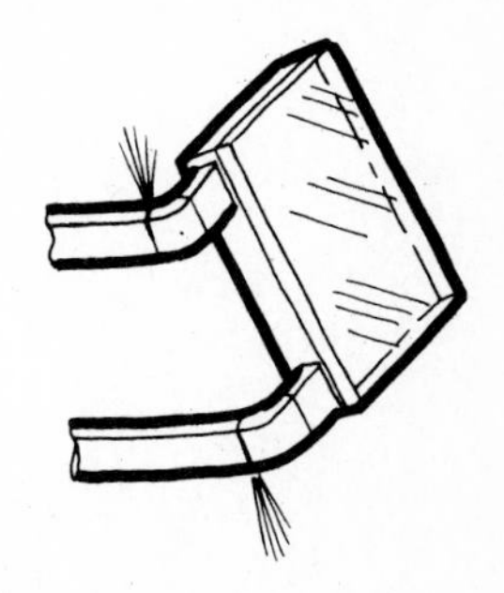

collectors and pipes or by freezing, as we'll see later. Leaks in air-heating systems won't ruin your rug, but they can mean a big loss in efficiency. Unless air leaks in ducts and collectors are carefully sealed, heat can easily leak away. If there are air leaks, they are very hard to detect.

But the greatest advantage of an air-heating system is that no freezing or boiling precautions must be taken. Water in a hydronic system can freeze, rupturing a solar collector's water-carrying passages. Since almost everywhere in the United States except Hawaii, southern California, and southern Florida have occasional freezing weather, hydronic systems must have some type of freeze protection. Antifreeze is often added to the water as protection against freezing, or *drain-down* systems can be installed to drain the water from the collector at night to prevent it from freezing.

Water can also boil in a hydronic solar collector if it gets too hot. It probably won't boil while the pump is running, since the water pumped through the collector keeps it relatively cool. But if the pump breaks down or the electricity goes out, the collector can get *very* hot. Stagnation temperatures of 300° F and

Some Advantages of Air-Heating Systems

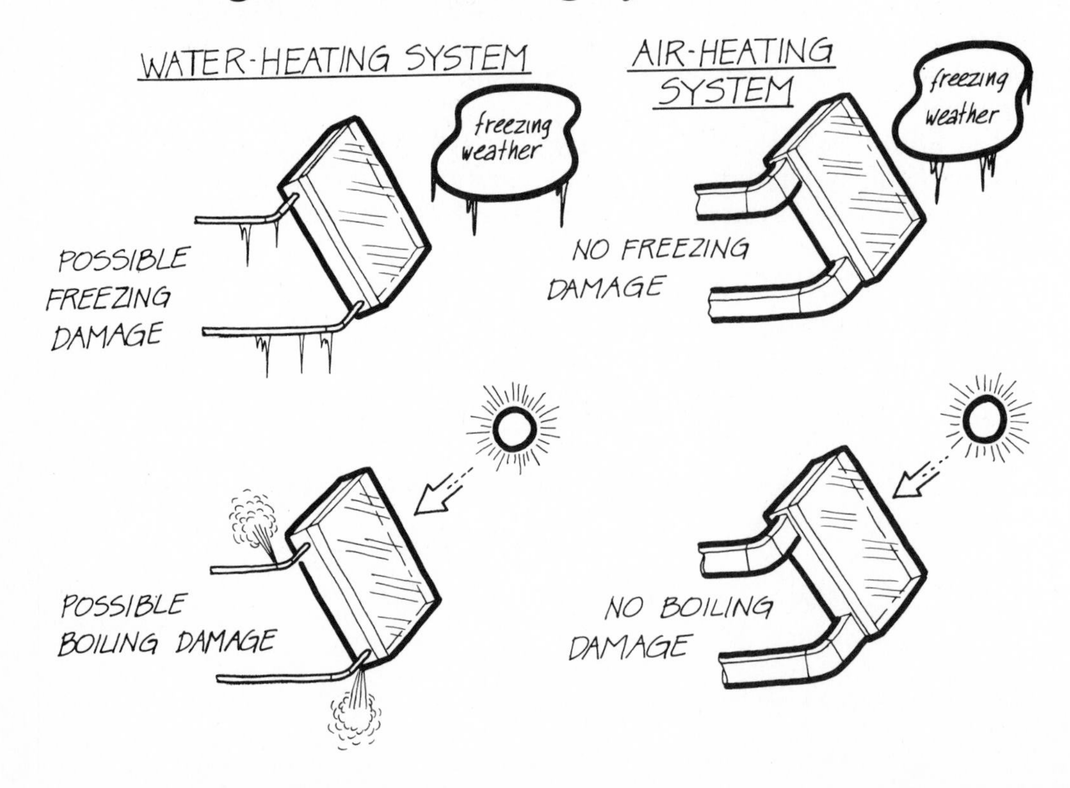

higher can occur on a bright, sunny day. To prevent the collector water from boiling, chemicals like glycol can be added, or the system can be set up to drain the collector automatically if the pump stops working. In any event, freezing and boiling precautions can add expense and complication to a solar heating system that an air-heating system doesn't entail.

What about the disadvantages of air-heating systems? There are some. First, an air-heating system tends to use more electricity driving its fans than a hydronic system does driving its pumps. Second, the ducts needed to transport the heat from the solar collector to the storage tank and then from storage to the rest of the house can be much bulkier than the piping in a hydronic system. Also, the bulky ducts are generally more expensive to install than are the pipes.

Some Disadvantages of Air-Heating Systems

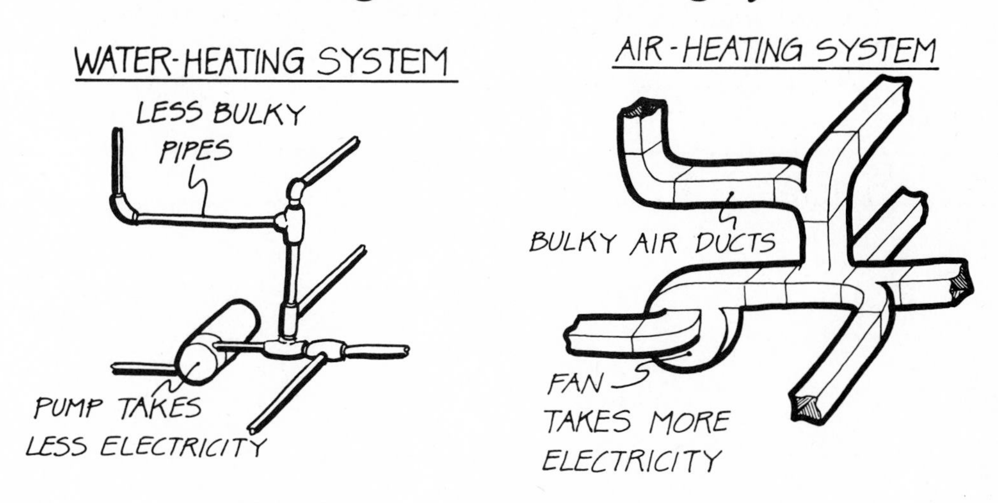

Now that we've looked at solar energy systems that use pumps or fans to transport heat from collector to storage, let's look at some systems that are truly solar powered.

A PASSIVE SYSTEM— DIRECT GAIN 20

Passive solar heating systems are those which don't use pumps or fans to transport the solar heat. They are passive because all they use to operate is the sun's heat. *Active* systems, such as we've learned about so far, use active power—usually electricity—to operate.

A *direct-gain* passive system is the simplest of passive solar heating systems. As with most passive systems, they are

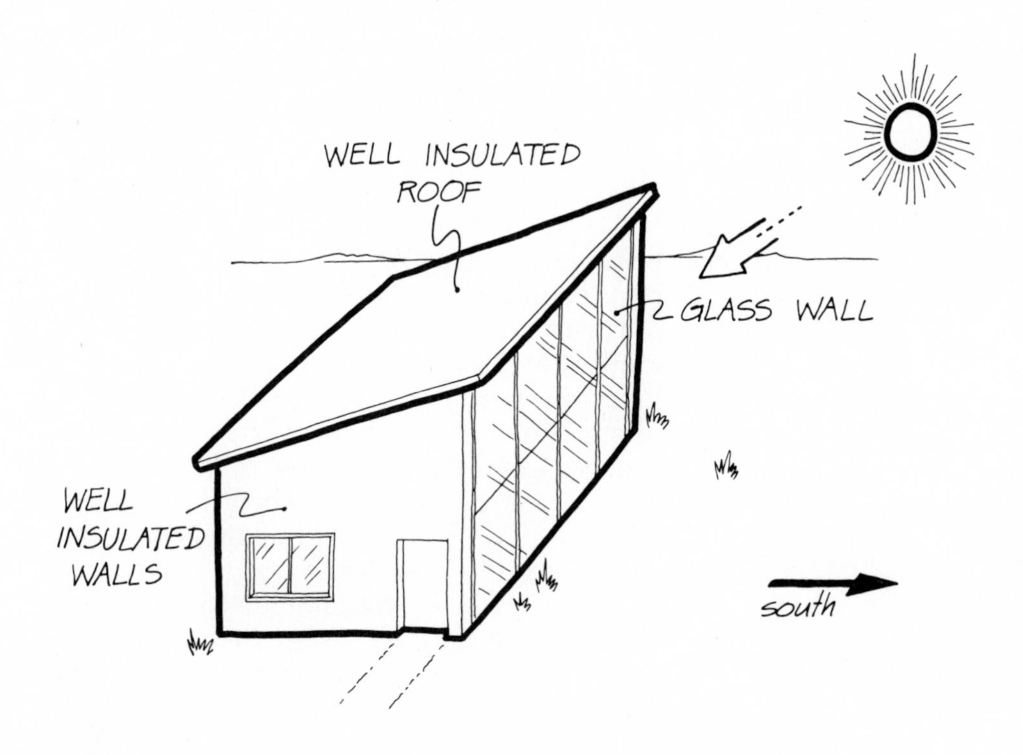

A Direct-Gain Passive Solar House

used primarily for house heating. In a direct-gain system, the house itself becomes a huge solar collector. Imagine a house that had the entire south wall made of glass and had lots of insulation on the roof and other walls. Sunlight entering the south-facing wall would be absorbed by the floor, furniture, and inside walls. The heat would be trapped inside by the glass wall just as a solar collector's heat is trapped by glazing. Well-insulated roof and walls prevents heat loss out of the house just the way a collector's insulation prevents heat loss out of the back of the collector.

A Direct-Gain House is Like a Solar Collector

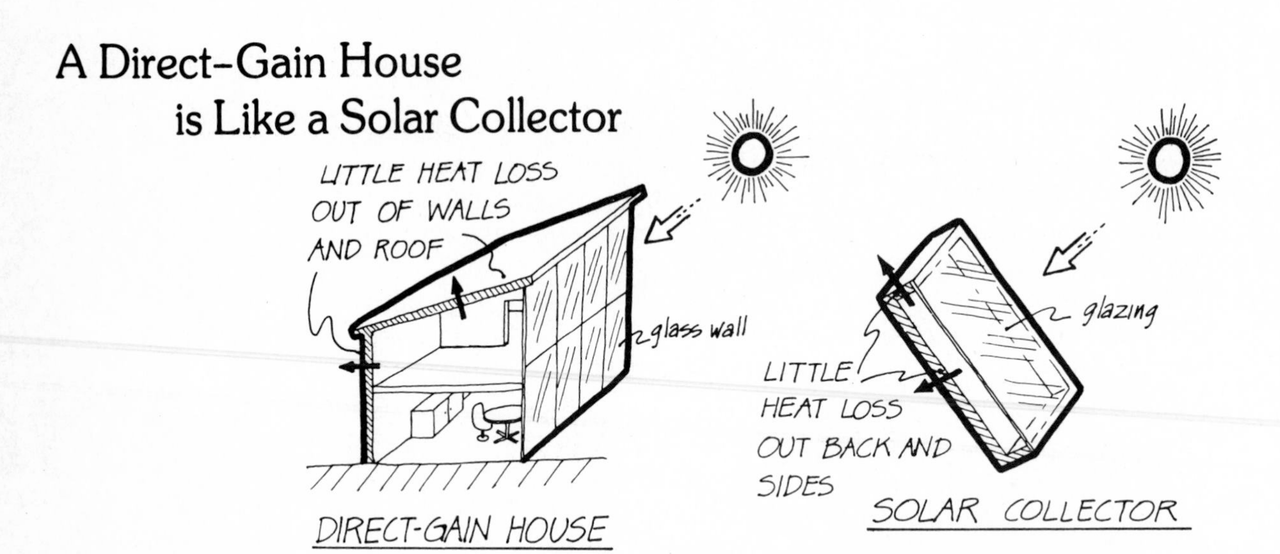

The simple direct-gain house might not be very comfortable to live in. During the day it could get very hot inside—remember how hot your car gets after sitting all day in the sun with the windows shut. At night it could get cold, because little of the solar heat collected during the day would be available to heat the house at night. For some buildings these problems aren't crucial. A school, for example, needs heat only during the day, and windows can always be opened to keep the building from getting too hot. Homes and most other buildings, though, need heat at night as well as during the day.

Active solar heating systems need storage to hold solar heat until night, and so do passive systems. Storage can be added to the direct-gain system just described by constructing the floor and inside walls of material that stores heat. For example, the floors can be made of thick rock or concrete. When the sun strikes the floor, the rock heats up and remains warm until nighttime; as it slowly loses its heat, it keeps the house warm.

Heat Storage in a Direct-Gain House

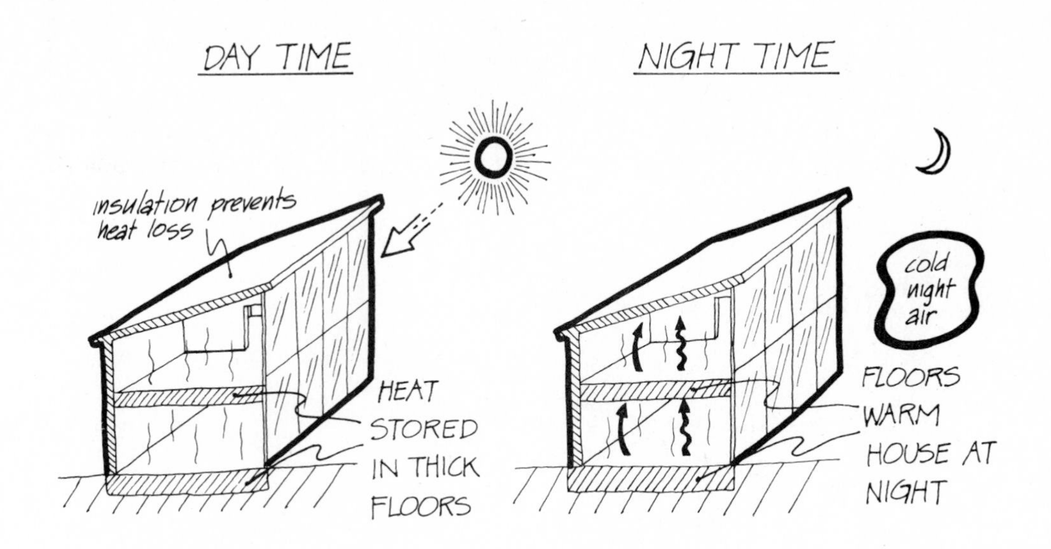

Sometimes drums of water are put inside a room to increase heat storage—water, remember, stores more heat than rock or concrete do. For best results, the water drums or thick rock floor should be in the direct sunlight, so that the storage is also the solar radiation absorber. The more heat storage there is in water drums and rock floors, the less tendency there is for a direct-gain house to overheat. A huge heat store will not get as hot after a day of absorbing solar heat, nor will it cool as quickly at night. The bigger the heat store, the less the *temperature swing*—the tendency to be too hot in the late afternoon and too cold in the morning.

The rainwater analogy to a direct-gain house is a collector tray directly attached to a large storage tank. The tank is large because the equivalent heat storage must be large to reduce temperature swing. The tank represents the heat stored in the house's floor or water drums; the water depth in the tank represents the house's temperature. A house loses heat to the outdoors; analogously, the tank must also have a leak. As before, the pipe on the collecting tray has a leak; it represents the heat leaking from the glass south wall on our direct-gain house. The rainwater tray collects the rain, filling the tank. When the rain stops, no rainwater flows down the pipe but water is still lost from both the pipe leak and the tank leak. The water level will slowly drop as the rainwater leaks out. It takes a long

Analogy of a Direct-Gain Solar House

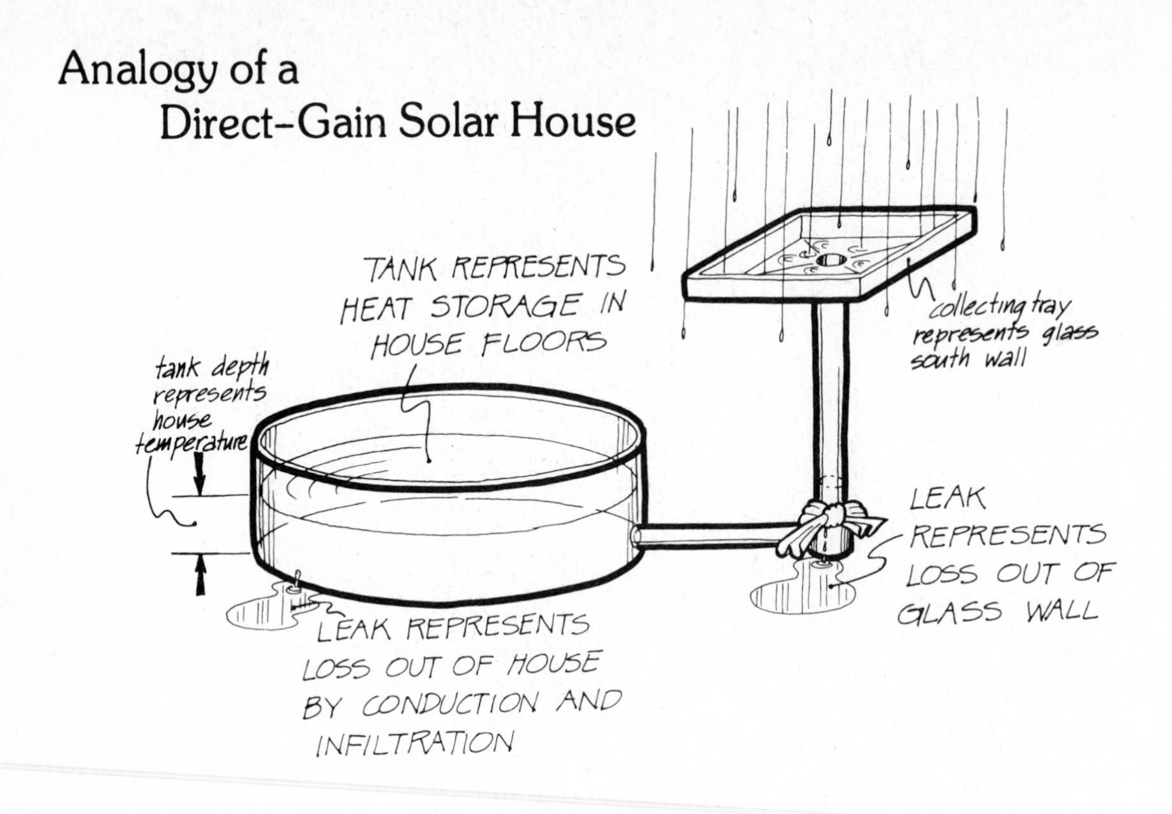

Water Level in Large Tank Changes Little.

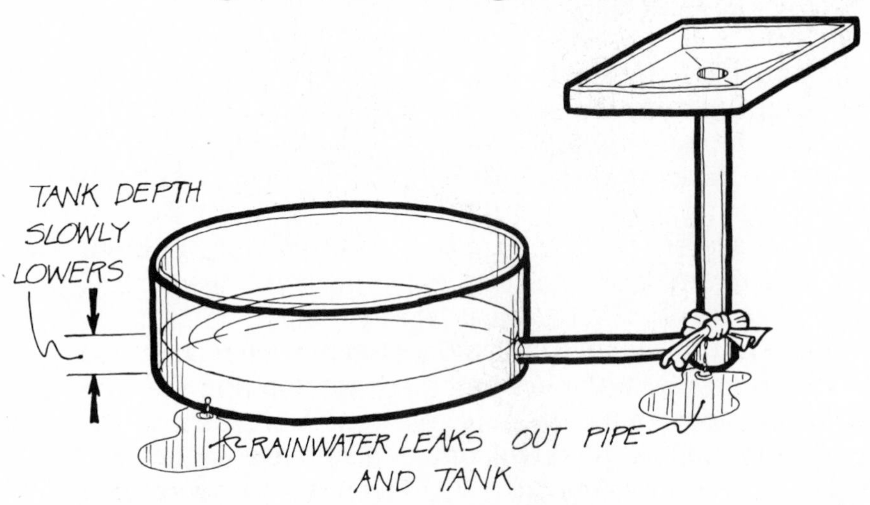

time to fill a large tank, so its depth doesn't rise very high; nor does a large tank drain quickly. The bigger the tank is, the less is the rise and fall of the water depth. Analogously, the bigger the heat storage is, the less the temperature swing is.

When we learned about active systems, we learned that the controller serves an important function in preventing nighttime heat loss. The solar collectors in an active system are separated from the storage. Only the heat transfer medium (air or water) moves heat from the collectors to the storage. When the controller turns off the fan or pump, no heat can be transported from storage back to the collecor. In a direct-gain system, though, the collector and storage aren't separated. The solar collector *is* the heat storage (the thick concrete floors).

Collector and Storage are Separated in an Active System

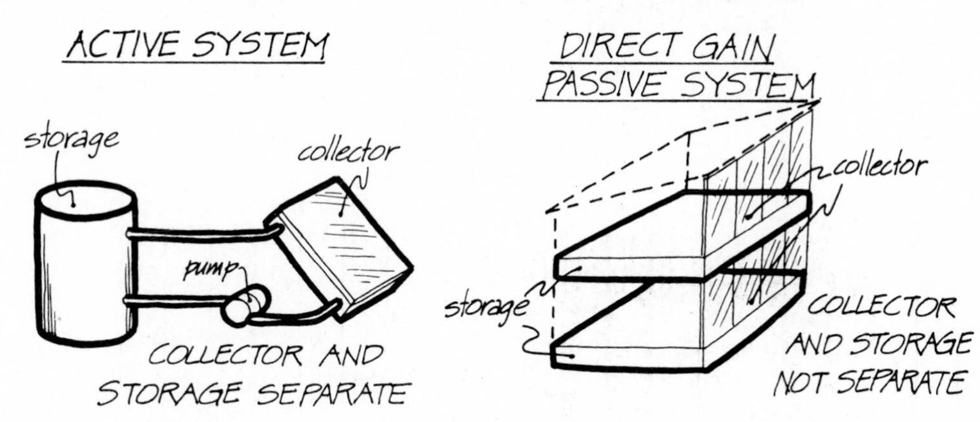

Unless the collector and the storage are separated, heat can leak from storage out the collector's glazing. Remember, in Chapter 8 we found that windows leak ten times the amount of heat of a well-insulated wall. An entire south wall of glass, then, represents a big heat loss. Even though the glass wall admits solar radiation during the day that heats the thick floor, at night the same glass wall loses much of the stored heat to the outdoors.

Not only do active solar heating systems have their heat storage separated from their collectors, but their storage is also separated from the living space. Stored heat can be used *when required:* a heat exchanger can extract solar heat from a water storage tank in a water-heating system, and in an air-heating system a distribution fan can blow solar-heated air from a rockbed. More important, when heat isn't needed, it can remain in storage. But in a direct-gain passive house, the heat storage isn't separated from the living space. If the floor of your direct-gain house gets too hot or too cold, so do you.

Storage and Living Space are Separated in an Active System

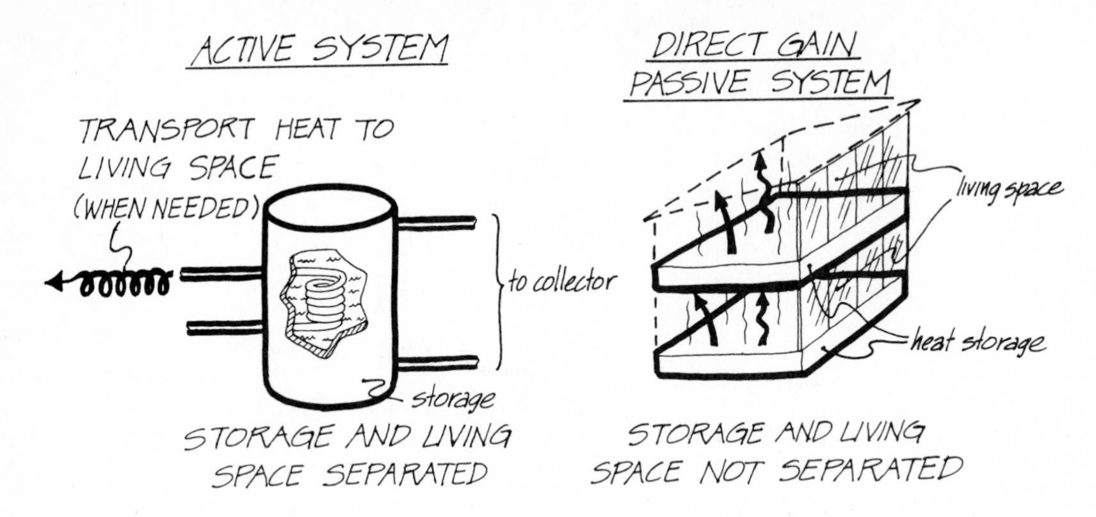

As we look at other types of passive solar heating besides the direct-gain system, we'll think of them as having either a *diode function* or a *control function*. Diode is a term borrowed from electronics, and it refers to a device that lets electric current flow in only one direction. Thus, the diode function of a passive solar heating system lets heat flow in only one direction—into the house but not out of it. The control function, on the other hand, lets stored heat be controlled—distributing it to the living space when and where it's needed.

Active systems accomplish the diode function by separating the solar collectors from the storage. The two are thermally linked only when it's advantageous: when the collector is hotter than the storage. An active system accomplishes the control function by separating the storage from the living space. The two are thermally linked only when it's advantageous: when the living space gets too cold.

PASSIVE SOLAR: THE DIODE FUNCTION 21

The diode function—letting solar heat flow in but not letting it flow out easily—can be added to direct-gain solar heating by *movable insulation*. To do this, insulation is moved in front of the glass south wall at night. One passive heating system, called *Drum-wall,* has a thick insulating wall that hinges at the ground. At night it is pulled up against the glass to keep heat from leaking out; during the day it is lowered to let solar heat through the glass wall. The system is called Drum-wall because just inside the glass wall is a stack of water-filled drums used for heat storage.

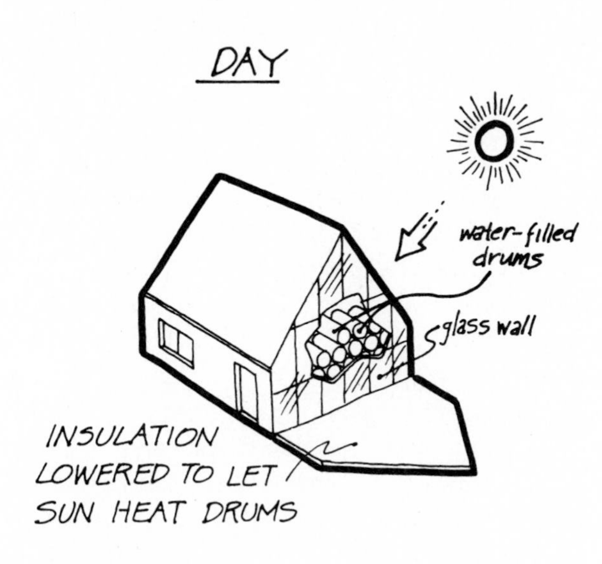

Movable Insulation Prevents Nighttime Heat Loss

The south wall of a house isn't the only place where the diode function can be utilized. Another passive design called *Skytherm* uses the roof of a flat house both as a solar collector and as a storage medium. The system stores heat in water-filled plastic bags, similar in construction to water beds. The bags rest on a metal roof so heat from them can be conducted through the roof and then radiated to the living space below. The diode function works by means of large slabs of insulation that slide on tracks over the bags, insulating them at night. The black-colored bags absorb solar heat during the day when the insulating slabs are moved aside by a device similar to a garage-door opener.

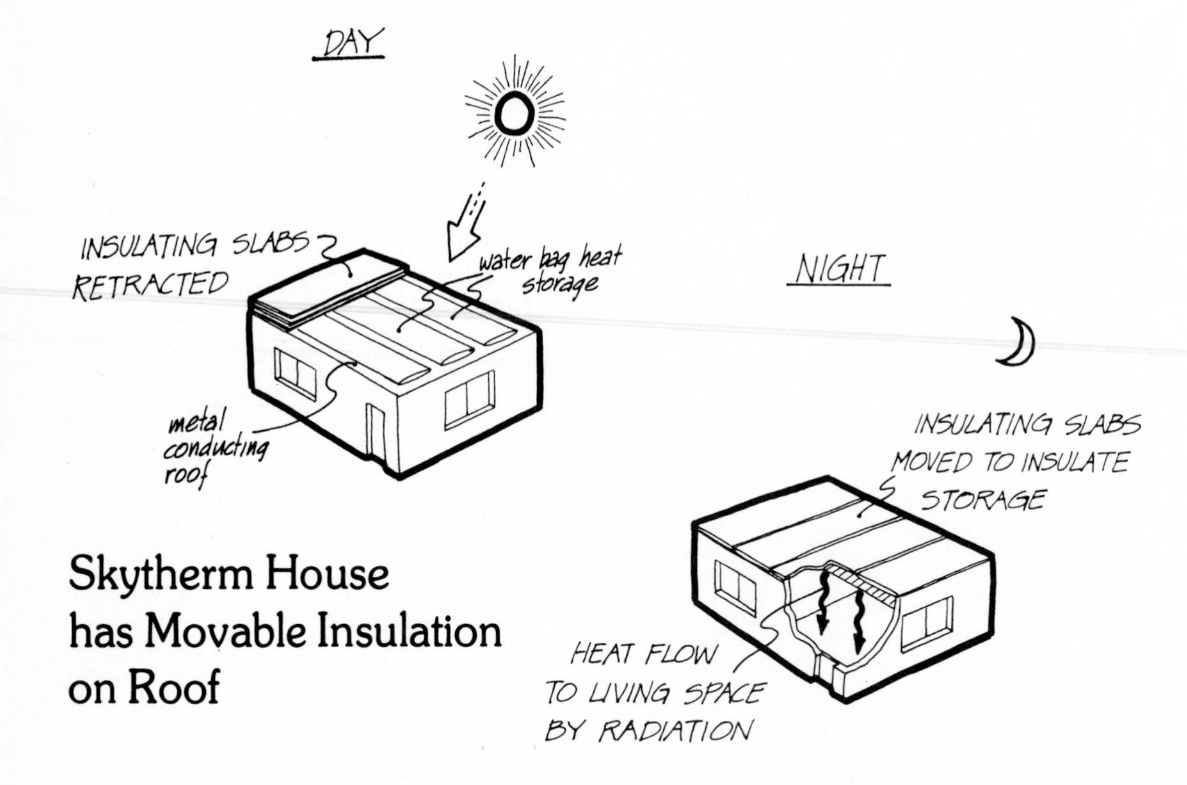

Skytherm House has Movable Insulation on Roof

Movable insulation doesn't always mean moving big slabs of insulation. In *Beadwall*, another type of passive solar heating, the diode function works by moving Styrofoam beads. A fan blows the beads into the space between two glass layers on the south wall of a house. When in place, they provide excellent insulation against heat loss through the glass. When the beads are removed by a suction fan, solar radiation goes through both glass layers and heats the storage within. Sometimes large vertical fiberglass or metal tubes filled with water store heat in a Beadwall system.

Foam Beads Prevent Nighttime Heat Loss

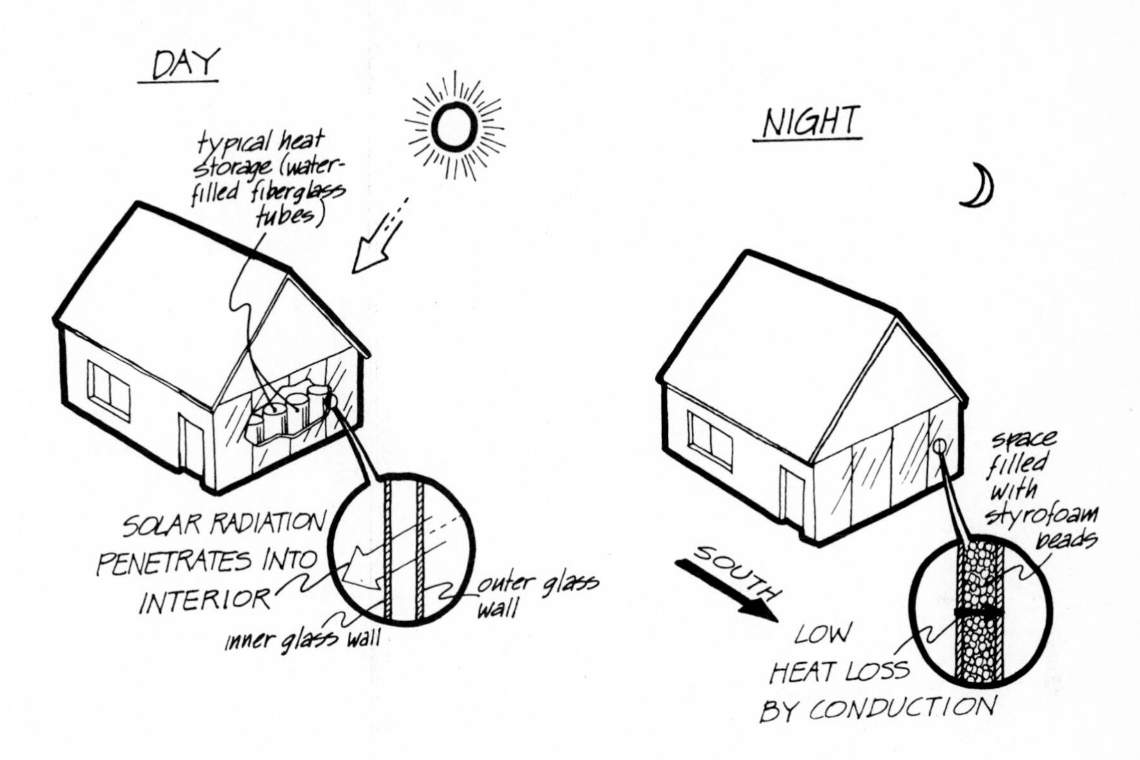

Not all passive heating systems use movable insulation to achieve the diode function. A system called the *Trombé wall* (named after its French inventor) uses quite a different technique. A masonry wall is built just inside the glass south wall of a direct-gain house leaving a gap a few inches wide between the masonry and the glass. This gap is connected to the living space by holes in the masonry near the floor and the ceiling.

Air in the gap between the glass and masonry is heated by convection from the black-painted wall. The lighter, heated air tends to rise, just as hot air in a chimney rises. It flows out through the upper holes and is replaced by cool air flowing in through the bottom holes. This cool air is, in turn, heated by the wall and flows out through the top holes. Of course, the cool air flowing in the bottom is the previously heated air which has given up its heat to the room. The process of circulation based solely on heating and cooling is called *thermosyphon* circulation. Instead of a fan blowing air from the solar collector into the living space, the air moves there by itself. Heat moves by transport heat flow, in a process very similar to natural convection.

Thermosyphon Circulation of Trombē Wall

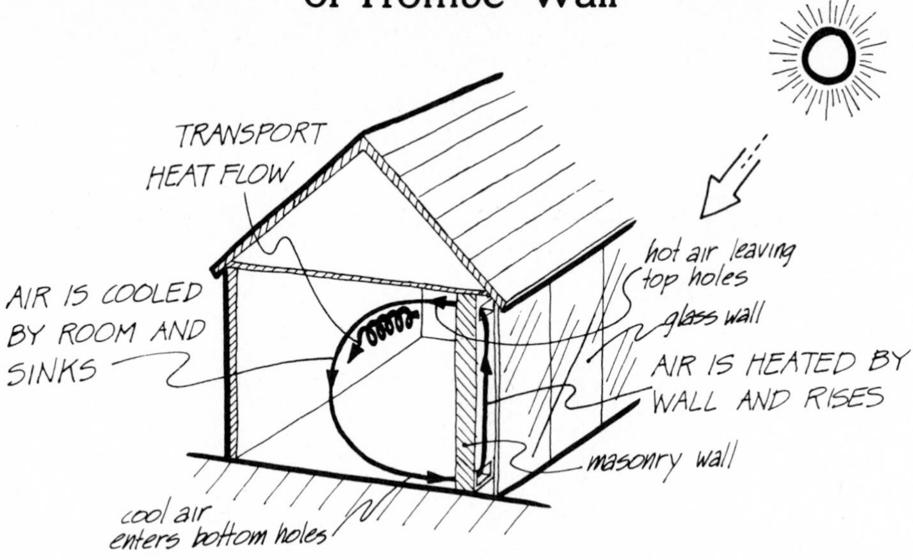

Heat can be *lost* by thermosyphon action at night. Air in the gap cooled by the cold night air tends to sink and flow out through the bottom holes and is replaced by warm air flowing in through the top holes. To prevent *reverse thermosyphon*, as it is called, the top holes are loosely covered by a thin plastic flap. The flap operates like the one in the check valve of our rainwater analogy—it lets warm air flow into the house during the day but won't let it flow back out again at night.

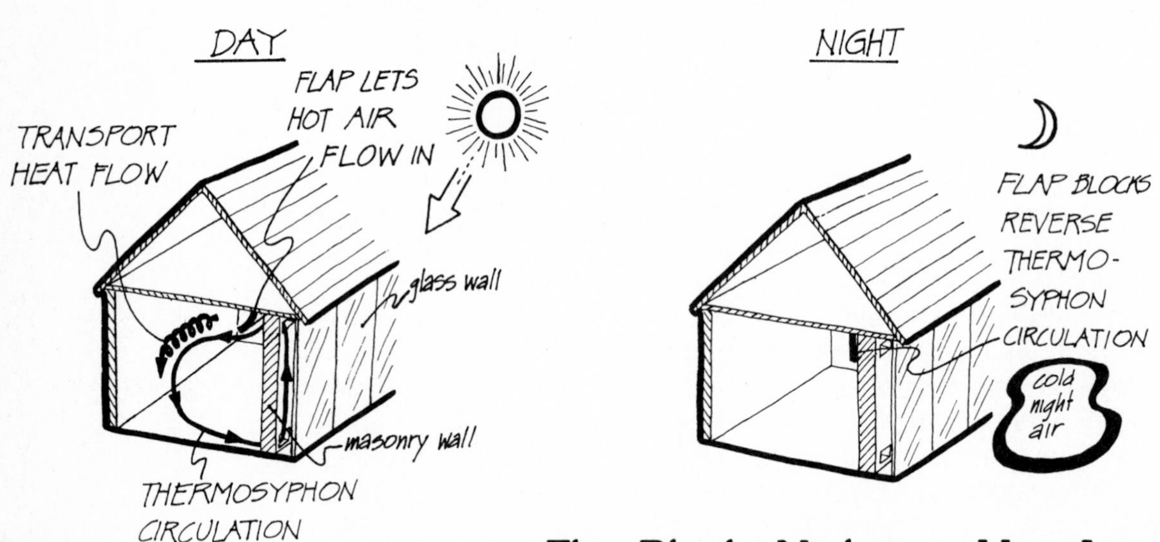

Flap Blocks Nighttime Heat Loss

The passive heating systems we've just discussed illustrate the diode function. They show how a direct-gain system can be modified to reduce nighttime heat loss. But that is not to say these same systems can't perform a control function as well. Next, we'll see how these and other passive systems can control solar heat delivered to the living space.

PASSIVE SOLAR: THE CONTROL FUNCTION 22

The control function in passive solar heating affects both when and where solar heat is delivered to the living space. *When* considers the timing aspects. Heat delivered when a house is cold is better than heat delivered when the house is already warm enough. *Where* considers distribution. Heat must be delivered to all rooms in a house, not just the ones that happen to have a southern exposure.

One aspect of the control function is *seasonal*. For example, the movable-insulation methods (Drumwall, Skytherm, and Beadwall) can also be used to adjust the delivery of solar heat based on the season. Overheating can be reduced by varying the amount of time during the day that the insulation is in place. For example, a Skytherm house might have its heat store collecting solar heat all day in the winter, half a day in the spring, and not at all in the summer.

Seasonal control of solar heat can also be achieved by using awnings and trees, shrubs, or other plants. Awnings and leafy plants shade a direct-gain house in summer when heat isn't needed but let solar heat in during the winter. Since the sun is high in the sky in the summer, an awning generally provides more shade in the summer than in the winter. Leafy trees don't have leaves in the colder months, when solar heat is needed.

Daily control of solar heat is another aspect of the control function. One way you've already learned about to control heat on a daily basis is to use lots of storage. If storage is very large, the house never gets too hot during the day yet it can warm a house all night long. The Drumwall, Skytherm, and Beadwall systems use huge amounts of water storage to reduce the temperature swing and are still able to provide heat overnight.

The masonry wall of the Trombé wall system achieves daily control in another way. The outer surface of the wall gets

Plants and Awnings Control Solar Heating

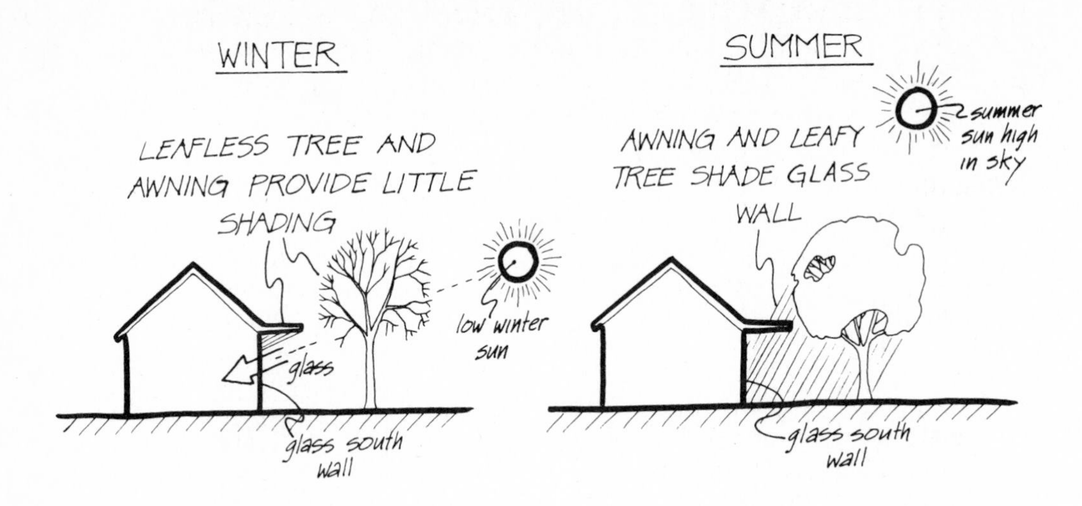

quite hot, but it is separated from the living space by the wall itself. As the wall heats up during the day, heat is conducted through the wall (in addition to the thermosyphon heating mentioned earlier). By nighttime, the wall has become quite warm and continues to conduct its stored heat to the living space. The

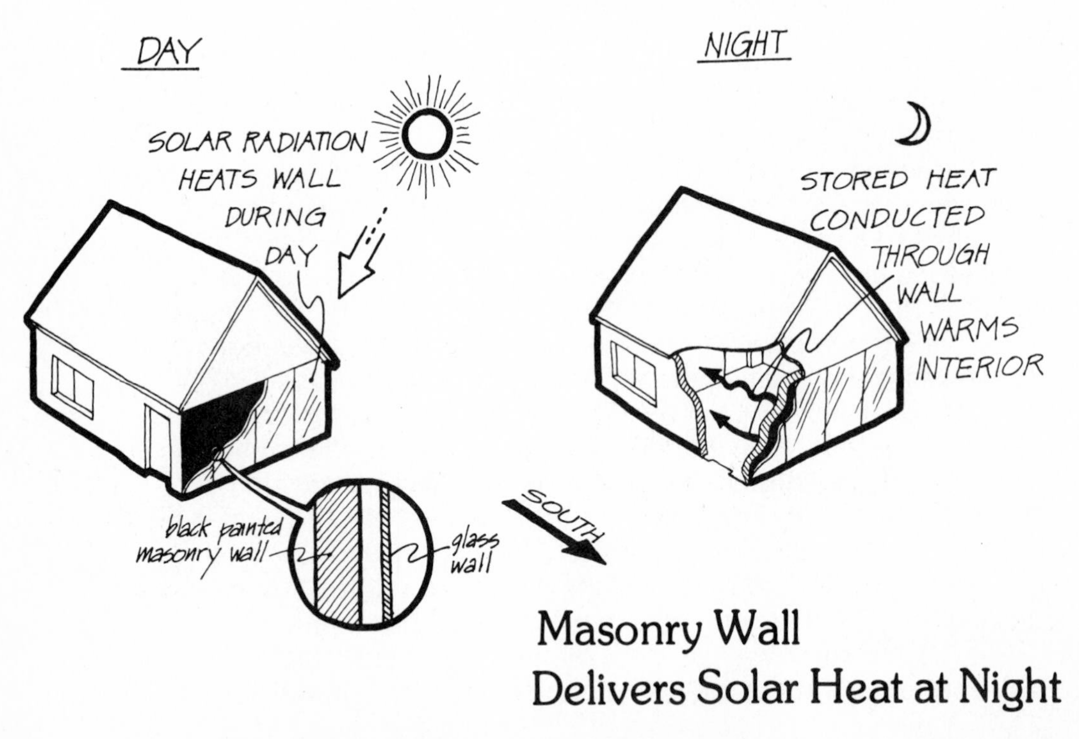

Masonry Wall
Delivers Solar Heat at Night

heat delivered by the wall to the living space is delayed from the late afternoon (when it's not really needed) until nighttime (when it is needed). This delay in delivering heat depends on how thick the wall is—a foot-thick wall seems to be about the best. Once the heat gets to the inner wall surface it is transferred to the living space by radiation and convection heat flow.

The control function may also be achieved on a daily and a seasonal basis by stopping the solar radiation from entering a direct-gain house. The *Sun Window* system, as it is called, replaces the glass south wall with a wall of clear plastic which has many passages molded into it. When the temperature inside the house gets too high, a dark-colored liquid is pumped through the passages. The passages are so close to each other that most of the sunlight that would enter the house is blocked. The effect is similar to lowering a window shade on the south wall—except that only the dark liquid moves anywhere.

Since the liquid is dark, it absorbs solar radiation. Not only does it keep the sun from entering a too hot house, but the fluid itself gets hot. The hot dark liquid is pumped to a storage tank where it can be used to heat water or the house later at night. The Sun Window system combines the simplicity of direct-gain passive heating with the control function of active solar heating.

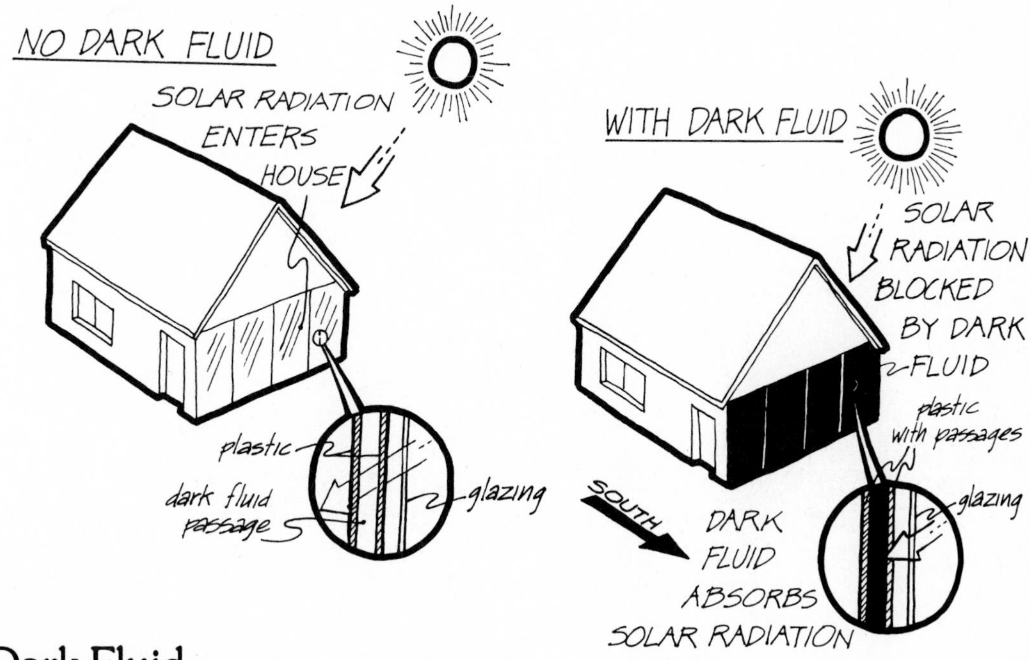

Dark Fluid
Controls Solar Heating

Passive systems can also accomplish the control function by separating the heat storage from the living space. The simplest approach to separating the two is to have a direct-gain greenhouse separated from the house itself. The greenhouse not only houses plants and stores heat but also acts as a buffer between the living space and the direct-gain system. In the late afternoon the direct-gain greenhouse might get quite warm, but that doesn't necessarily mean the living space will get too warm. A small fan can be used to draw in air from the greenhouse to warm the house when required, or vents can be opened to let the warm air flow into the living space when required.

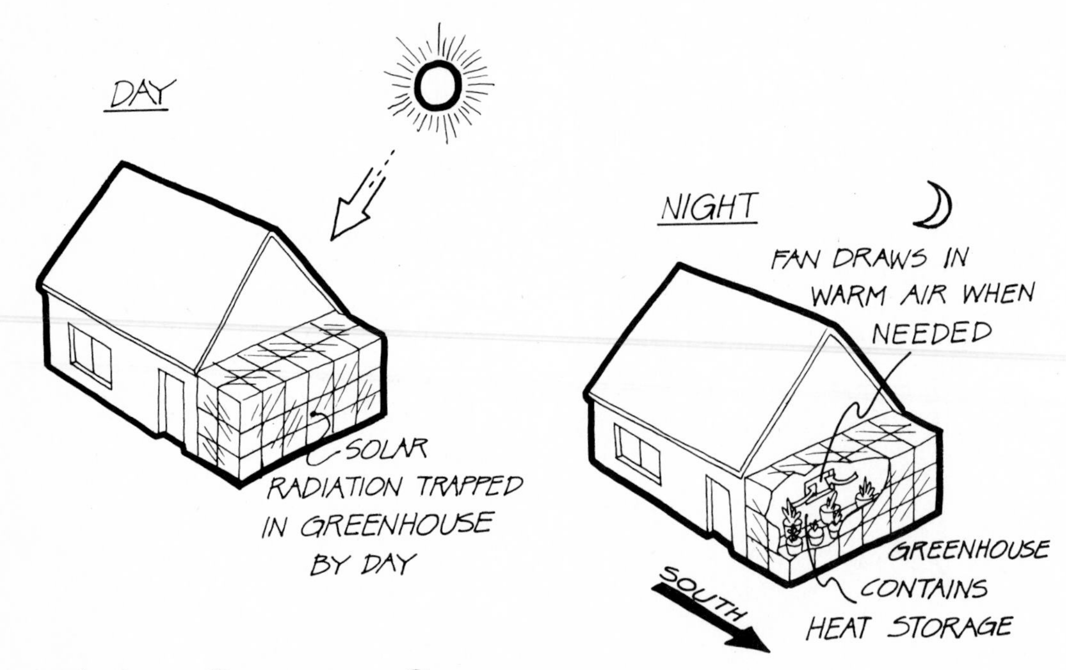

Greenhouse Separates Storage from Living Space

You may have noticed that some of the passive systems we've been discussing aren't strictly passive. The Sun Window method uses electricity to pump dark liquid through the plastic wall, and the greenhouse method can use electricity to blow heated air into the house. In fact, many passively heated houses use active methods when convenient. For example, passive solar houses often use fans to distribute heat from warm south-facing rooms to cooler north-facing rooms. As improved diode and control functions are added to the passive direct-gain house, the distinction between active and passive may become blurred. But after all, the purpose of any system is to provide solar heat as cheaply, reliably, and conveniently as possible, even if it means combining aspects of both active and passive solar heating.

THE THERMIC DIODE 23

The biggest advantage of passive solar heating systems is their simplicity and reliability. Since they don't depend on fans or pumps which wear out or controllers which malfunction, there's less chance of something going wrong. Considering how easy it is for homeowners to forget to change the filters in their furnaces, the simplicity of a passive system is appealing. But passive solar homes must usually be built from the ground up. It is expensive to add, say, a Trombé wall to your house after it has been built. However, some passive systems are suitable for *retrofitting*—for adding to a home that has already been built. One such system is called the *Thermic Diode*. The term Thermic Diode comes from its inherent diode function: it stores heat when the sun shines on it but doesn't lose heat at night.

The Thermic Diode is a *modular* passive heating unit. Modular means that it comes in individual units, or modules,

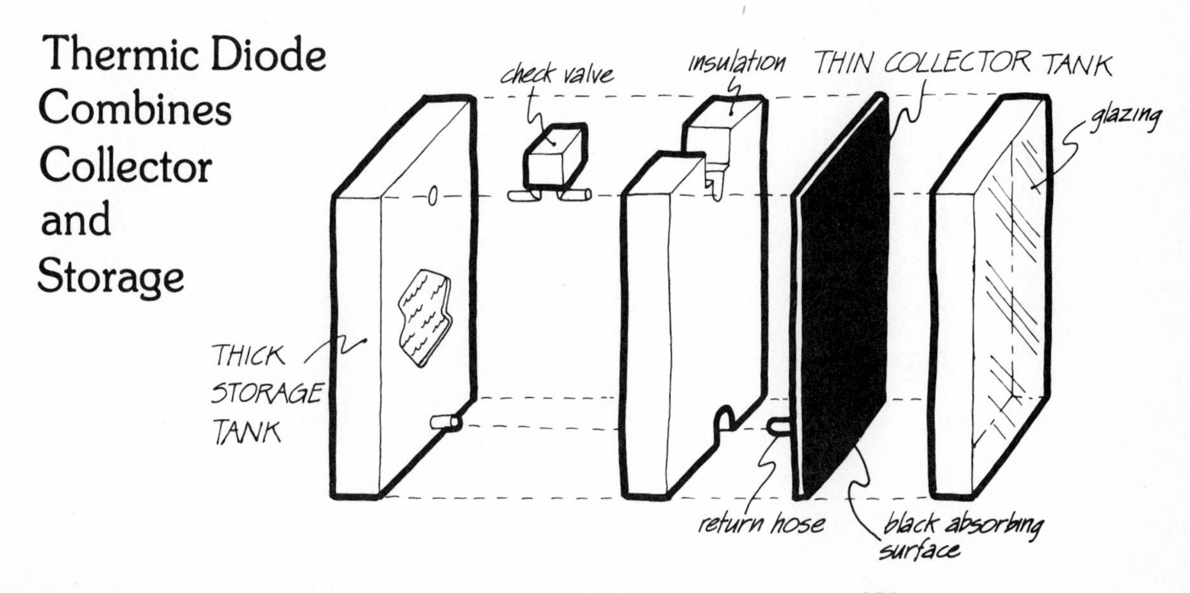

Thermic Diode Combines Collector and Storage

which can be mass-produced in a factory as the collectors of active solar heating systems are. But it is also passive; it has no fans or pumps and requires no electricity to operate. Each module contains a complete solar heating system: a solar collector, a controller, a heat storage unit, and even a heat exchanger. Yet the Thermic Diode is simple, rugged, and reliable. It doesn't even have any moving parts that could blind or corrode.

Functionally, the Thermic Diode consists of two flat tanks separated by insulation. The tanks are filled with water and are connected at the top by a check valve and at the bottom by a return hose. During the day, solar radiation passes through the glazing and is absorbed by the black outer surface of the thin collector tank. Heat is conducted through the tank's wall, heating the water inside. The heated water is less dense, so it rises and flows out through the check valve and into the storage tank. Cool water from the storage tank flows into the bottom of the collector by way of the return hose. As in a Trombe wall, the water circulates from the solar collector to the storage tank by the thermosyphon principle. At night, the collector cools and the water tries to thermosyphon in the opposite direction; but the check valve blocks this reverse circulation. The thick insulation prevents heat loss from storage when the valve is closed.

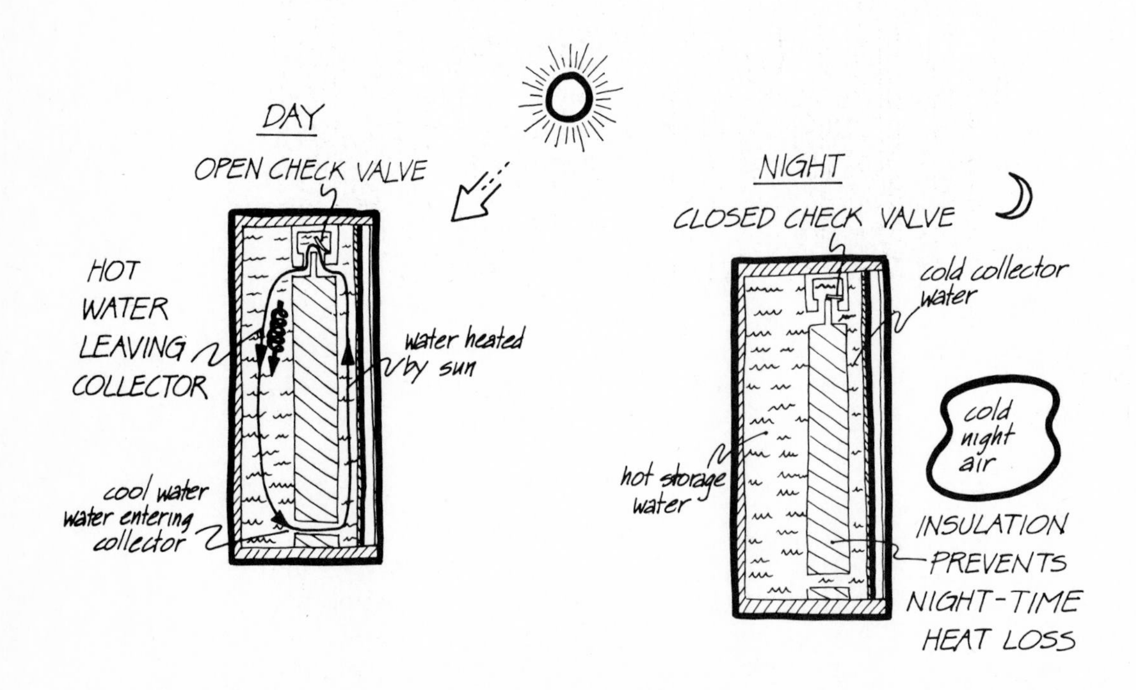

Check Valve Prevents Nighttime Heat Loss

The check valve can't be a simple flap valve like the one that we learned about in our rainwater analogy. The forces that circulate the water are too minute to open and close a flap valve, especially if the same valve has to work reliably for twenty or thirty years. The Thermic Diode uses a specially designed check valve that has no moving parts. The valve is simply a chamber filled with water with a layer of oil floating on the water's surface. The oil layer is thick enough so that it covers the top of a short tube coming from the collector. When the solar collector is hotter than the storage tank, the water in the collector tries to rise. It pushes up over the top of the tube through the oil; then it flows down through the outlet to the storage tank. But when the storage tank is hotter than the collector, hot storage water tries to rise and pushes oil down into the tube. Since oil is lighter than water, it resists being pushed down into the water—the way it is hard to push a beach ball underwater when you're playing at the beach. Oil is pushed only a short way down the tube; then it blocks the hot water trying to circulate from storage to the cold collector.

Floating Oil Acts as Check Valve

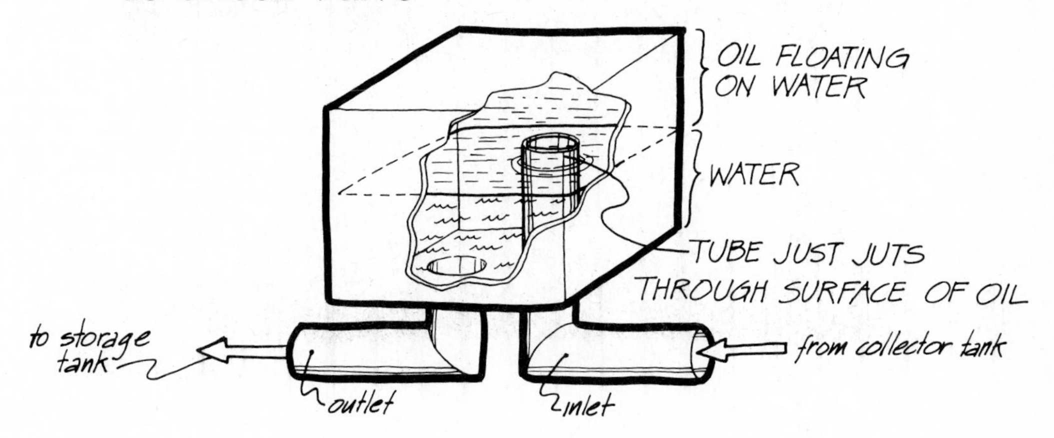

Functionally, the oil valve acts just like a check valve. It lets the hot water flow through in one direction but not in the reverse direction. Coupled with the thermosyphon circulation between the collector tank and the storage tank, the oil valve lets *heat* flow one way, but not the other. Thermosyphon circulation and the oil valve provide the same diode function that the controller and pump does in the active system.

Water Flows One Way Through Oil Valve

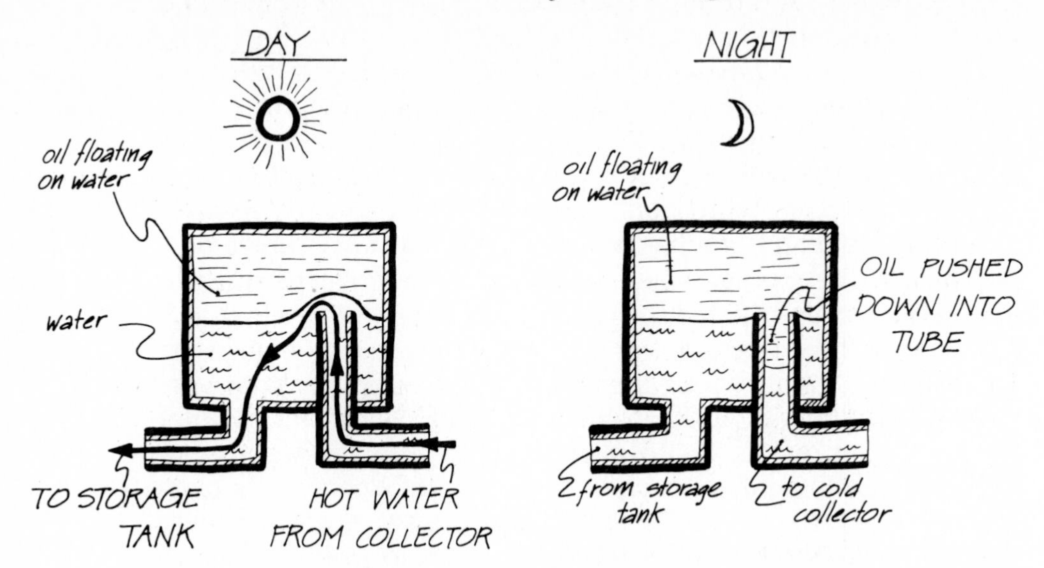

Heat stored in the storage tank can be used for hot-water heating or for house heating. In hot-water heating, a Thermic Diode has a heat exchanger built into the storage tank to extract heat. As we saw earlier for an active system, it can be used as a preheater for a conventional water heater. Only two pipes need be connected: one from the watermain, and one going to the conventional water heater.

Thermic Diode Water Heating

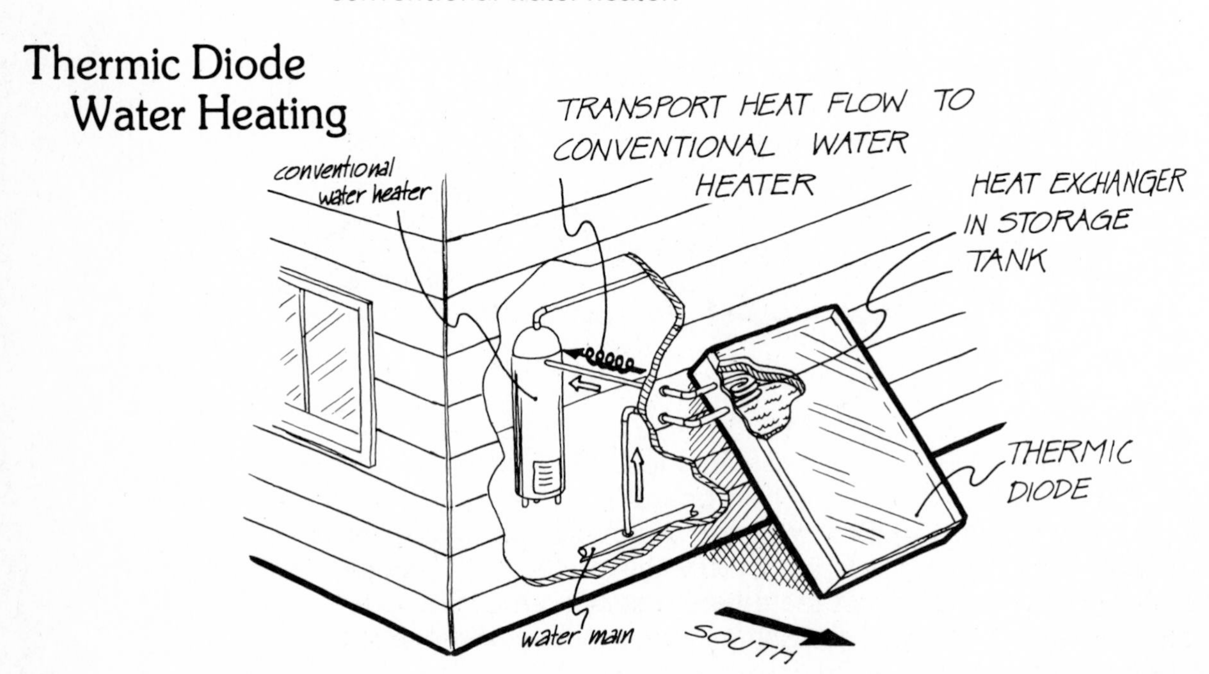

To heat a house by means of baseboard hot water, the heat-exchanger tubes of several Thermic Diodes can be strung together so that the output of one serves as the input to the next. The last tube goes to the auxiliary furnace. For house heating by forced air, the back wall of the storage tank itself can be used as a liquid-to-air heat exchanger. The back walls of several Thermic Diodes are exposed to a chamber formed between the Diodes and the house's wall. Heat flows by convection from the hot storage water through the storage tank's walls, and then by radiation and convection into the chamber. Since the storage tank is flat, lots of surface area is exposed to the chamber, letting the heat flow easily from the stored hot water. The heated chamber air is drawn into a conventional forced-air furnace for additional heating (if needed) before distribution to the living space.

Thermic Diode House Heating

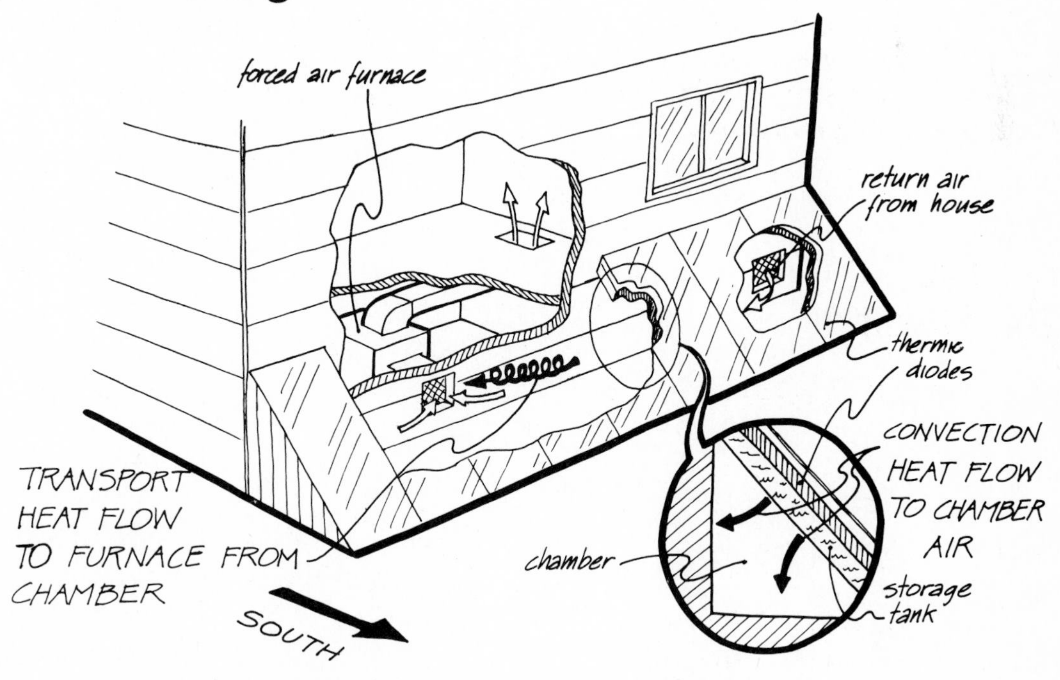

In either hot-water heating or house heating, this passive system separates storage from both the collector (reducing nighttime heat loss) and the living space (allowing heat to be delivered only when it's needed). It performs both the diode and the control functions.

REFLECTING SOLAR RADIATION 24

When we learned about solar radiation we found that it could be reflected by mirrors or light-colored surfaces. If we can reflect solar radiation in our solar collector, we can capture more sunlight. The more sunlight we capture, the more heat we can get out of the collector. This is true whether we are using an active or a passive solar heater.

The simplest way to reflect sunlight into a collector is from the ground. White or light-colored pebbles in front of a Trombé wall system can add one-third more heat. In cold climates, snow on the ground has the same effect, especially in winter when the sun is low in the sky.

Snow Reflects Solar Radiation

Ground reflection isn't too important in roof-mounted collectors, especially those with a low tilt angle. Reflectors—mirrorlike surfaces often made of plastic film with a shiny coating—are used to increase the solar radiation that normally falls on a roof-mounted collector. Reflectors are cheaper than collectors, so there's a cost advantage in adding reflectors to a system. A disadvantage is that the reflectors need to be adjusted regularly, though perhaps only from month to month. One system, called *Pyramidal Optic Condenser*, uses the roof peak of a house to hold reflectors. One of the reflectors is adjusted to capture both the low winter sun and the high summer sun.

Reflectors Capture More Sunlight

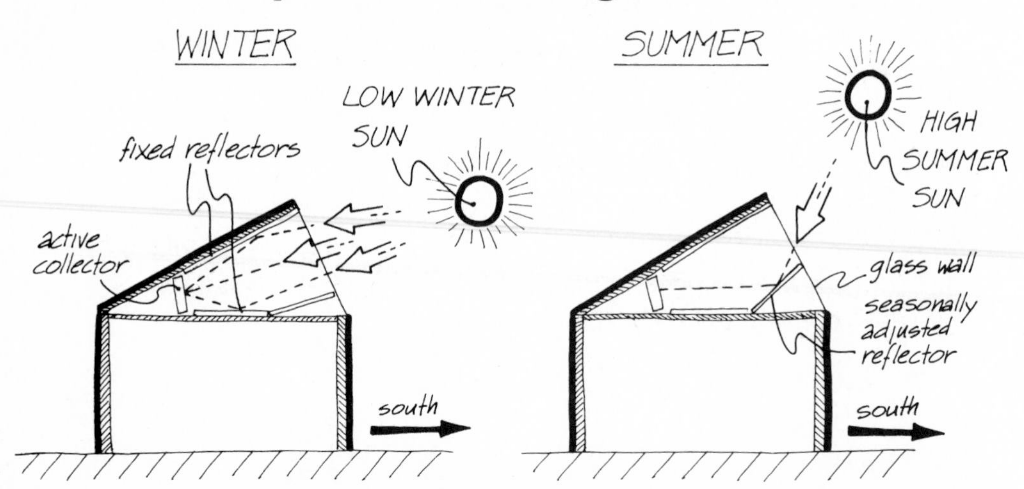

The rainwater analogy to a reflector is a sheet which catches some extra rain and lets it pour into the rainwater tray. Since more rain is captured, more will flow down the pipe. But even though we're capturing more rainwater, the leak doesn't get any bigger. Losses are the same with or without the "reflected" rainwater. Similarly, reflectors *add* heat (via solar radiation), but only collectors *lose* heat. In the system just discussed, the collectors cover only about half of the area normally needed to heat a house. Heat loss from the collectors is cut by half, yet the reflectors insure that the same amount of solar radiation is absorbed by the collectors.

Passive systems can also use reflectors. One method, called *Sun Louvers*, uses venetian blind louvers with a shiny upper surface to reflect sunlight onto the ceiling of a house or apartment. Ceiling tiles filled with phase change material absorb

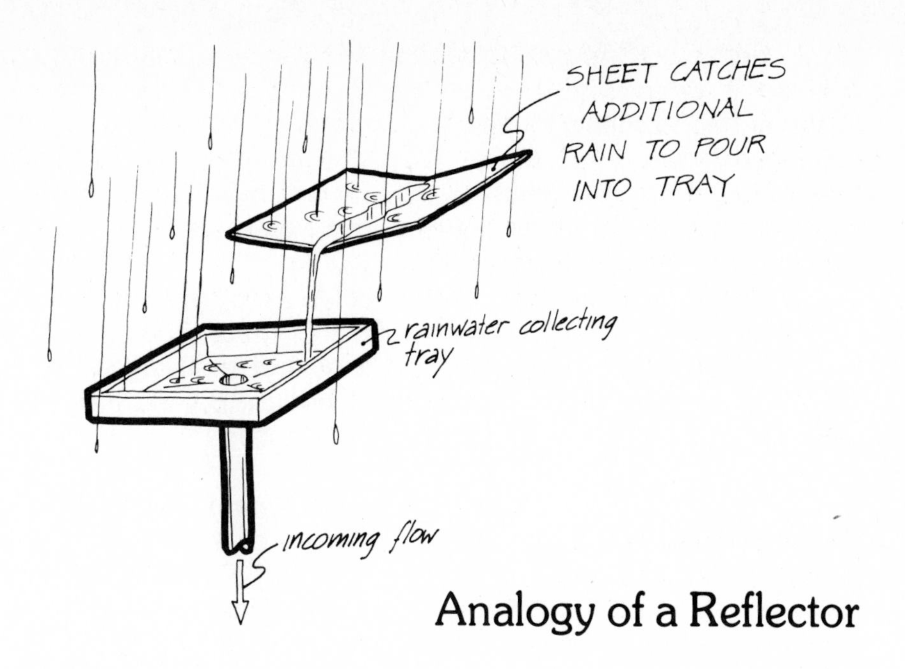

Analogy of a Reflector

the reflected solar radiation. As you may recall, phase change materials store heat without getting hot, so overheating is less of a problem. The phase change material melts as it absorbs solar radiation. Later at night, the melted phase change material solidifies and gives off heat.

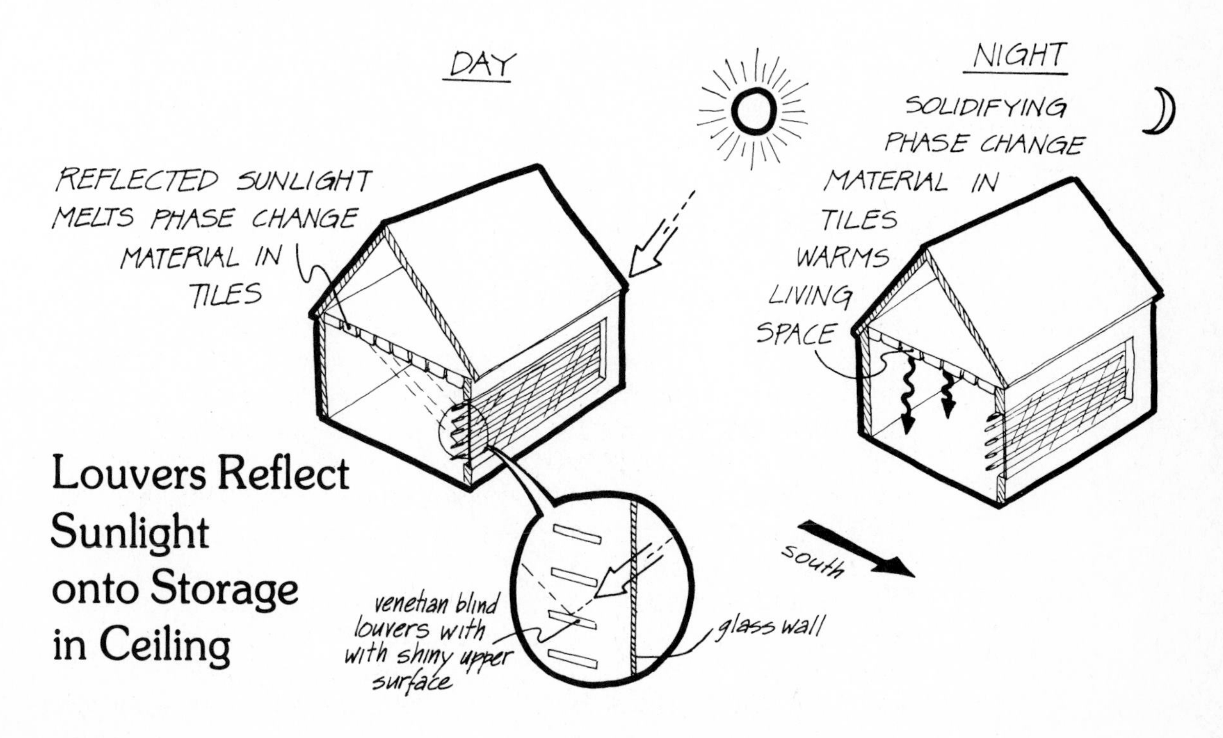

Louvers Reflect Sunlight onto Storage in Ceiling

Reflectors are also important in solar air-conditioning. It may seem strange that solar *heat* can be used for *cooling.* However, using thermodynamic principles, heat can be used to condense a vapor into a liquid. When the liquid evaporates again into a vapor it extracts heat—that is, it cools. You feel cold when you get out of the shower because the water evaporating from your skin has a cooling effect.

Most of the *thermodynamic cycles* used for cooling need heat at a very high temperature—200° F or more. The collectors we've discussed so far may get that hot, but their efficiency is very low. Ordinary glazing isn't good enough to yield high efficiency at these high temperatures; heat losses must be further reduced.

The collectors for solar air-conditioning, called *evacuated-tube collectors*, cut heat loss in two ways. First, the absorbing surfaces are surrounded by a tube-shaped vacuum jacket similar to a fluorescent light bulb. The vacuum jacket completely eliminates convection heat loss, since no air surrounds the absorbing surface. Second, the absorbing surfaces are made very small to reduce radiation heat loss—small areas lose less heat than large ones. Reflectors are placed behind the evacuated tubes, letting the smaller absorbing surfaces receive as much solar radiation as a larger surface would.

Heat is removed from the absorbing surfaces by conduction to tubes through which a liquid is pumped. The liquid often must be a special one that won't boil at a high temperature. So little heat is lost that these collectors achieve reasonably high efficiency even though they operate at very high temperatures; they are well suited to operating air-conditioning equipment.

Evacuated-Tube Collectors Use Reflectors

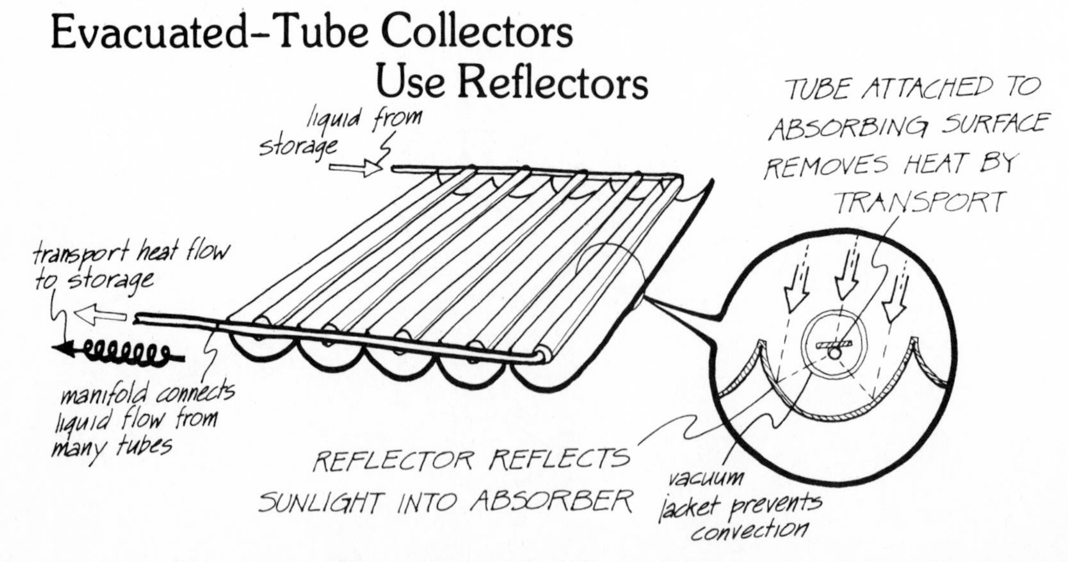

Using solar heat for air-conditioning should be a good match. The cooling is needed only when the sun is out; storing heat for nighttime use isn't necessary. However, the extra complication of reflectors, vacuum-jacketed collectors, and heat-driven air conditioners can offset the cost of storage. Nevertheless, evacuated-tube collectors can help cool buildings in many hot climates such as the Southwest; they are also effective in cooling many larger buildings such as stores in shopping centers.

WHICH SOLAR HEATER? 25

Now you've learned about all kinds of solar heating systems. Active water-heating systems have pumps that transport hot water and heat from a solar collector; heat is stored in a water tank. Active air-heating systems have fans that force air through ducts in the collectors and transport the heat to a rockbed heat store. In passive water-heating systems, the sun heats water without the use of pumps. For example, the Skytherm method uses movable insulation to prevent nighttime heat loss from solar-heated bags of water. Passive air-heating systems use the sun's energy to heat a house without using fans. The Trombé wall system is typical: thermosyphon action transports heat to the house by day, and a masonry wall delivers heat to the house at night.

Which of these systems is best? Which is best suited for the cold Northeast? for the sunny Southwest? While many factors are involved, perhaps the most important factor is economic: how much a certain method costs versus how much money it delivers. As you learned in Chapter 1, you pay an initial cost (sometimes called *first cost*) for a solar heater when it is installed, and you hope it will pay for itself, or even make money for you.

Imagine two empty jars on a balance scale. When you pay the initial cost of a solar heating system, it is like filling up one of the jars. The water in the jar represents your initial investment. You'd add water to represent the cost of the solar heater itself and the cost of installation (usually one-third of the total). Then you would take out the amount that represented the tax credits your state and federal governments give you to "go solar" (one-fourth or more of the total). At this point, you've invested a lot of money in a solar heater, and you've gotten no benefit from it at all.

Classification of Solar House Heating

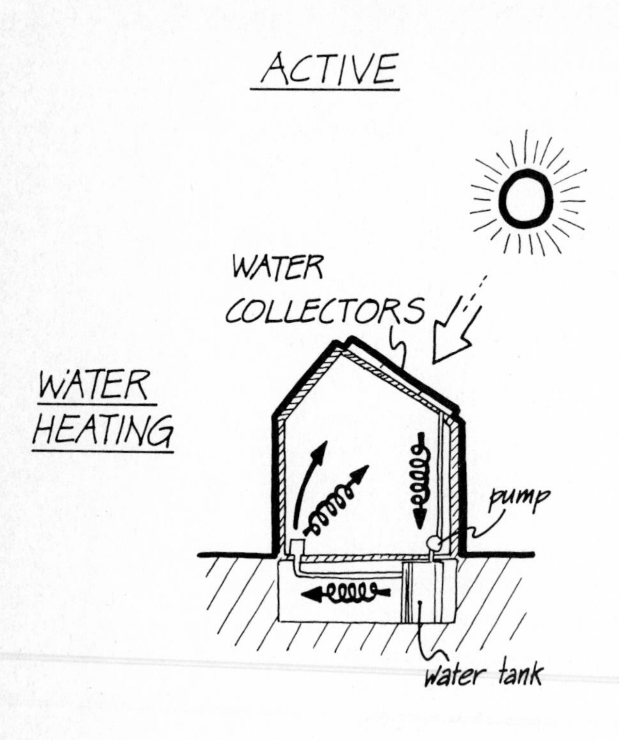

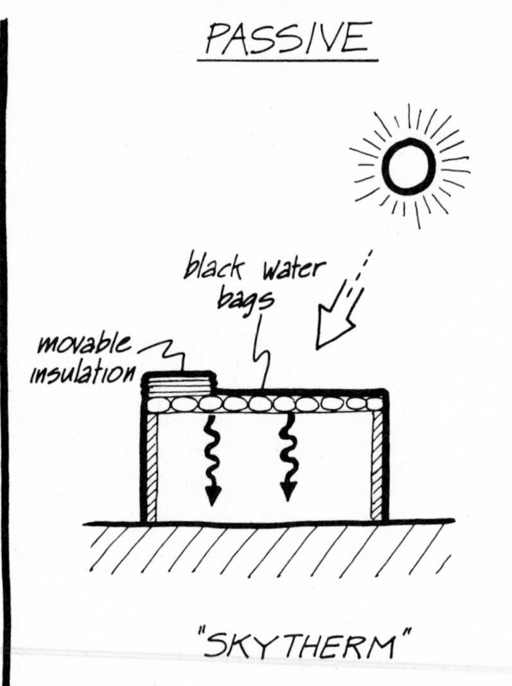

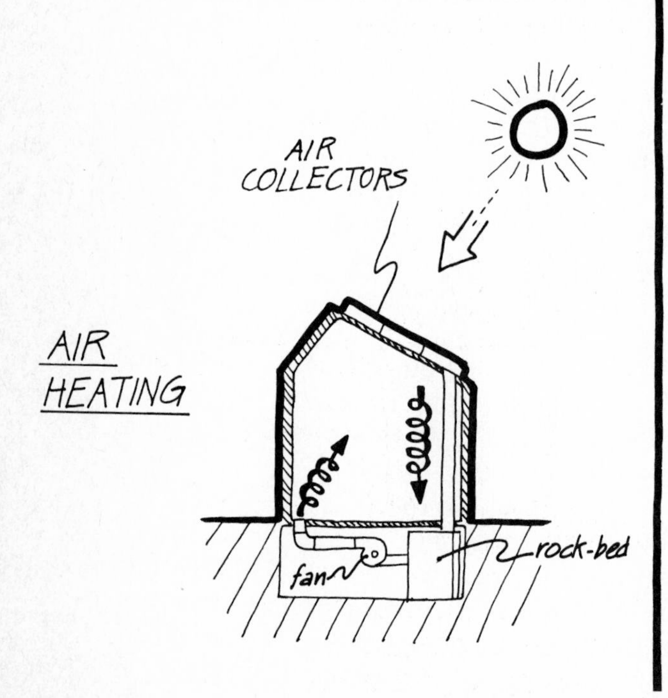

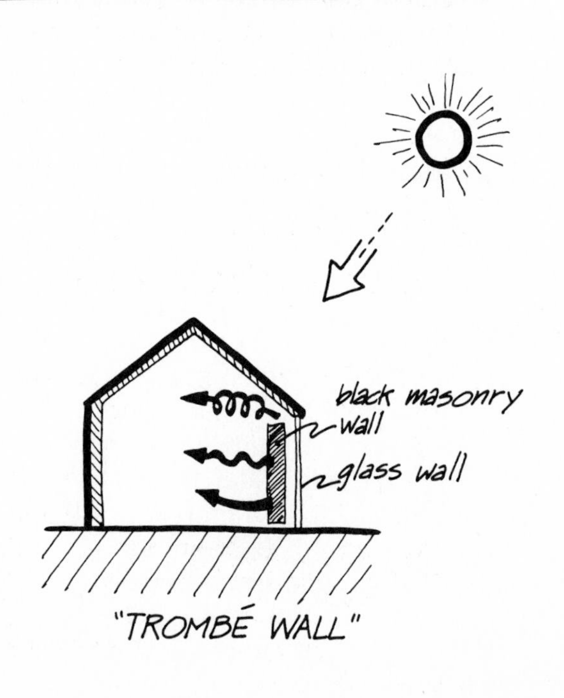

Large Initial Cost of Solar Heater

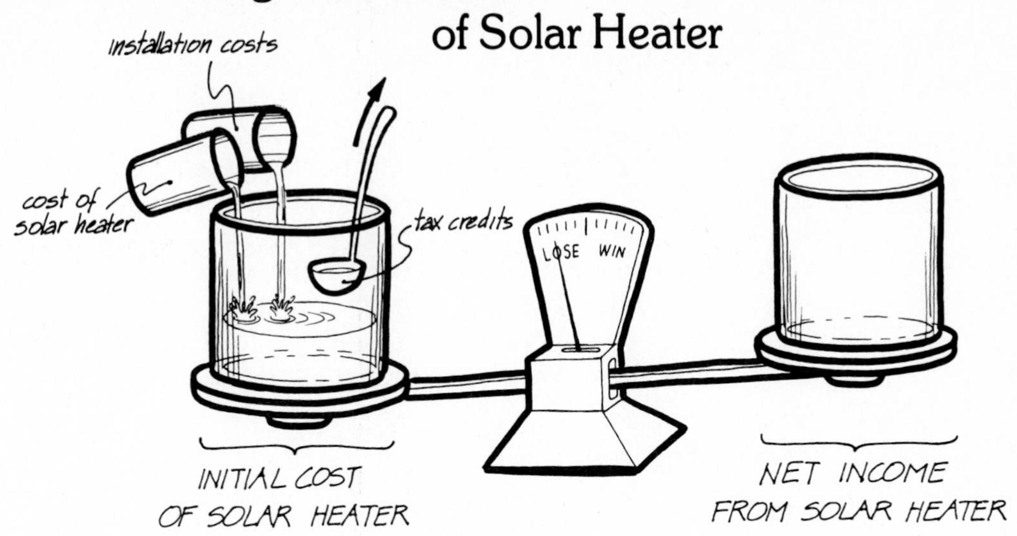

Over the next few years, however, you will save some money each year on your heating bills. The auxiliary unit won't be running all the time because some part of the heat you use will be solar heat. But you will also be paying out some money each year to maintain the solar heater in good operating condition, and possibly to pay interest on a loan you took out to buy it in the first place. We can add some water to the other jar each year to represent the net income from the solar heater: savings in fuel minus interest and cost of maintenance.

First Few Years: Initial Cost Outweighs Income

If the cost of fuel went up—as it seems to do each year—then each year your net income would be more. If the solar heater had a major malfunction and all the solar collectors had to be replaced, the net income would be much less. Either way the solar heater would be a loss, since the income over the last few years still wouldn't equal the initial cost.

After a few more years your solar heater would start making money. Each year you would add water to the net-income jar until the net-income jar outweighed the initial-cost jar. Thereafter you'd make money on your solar heater.

Several Years Later: Income Outweighs Initial Cost

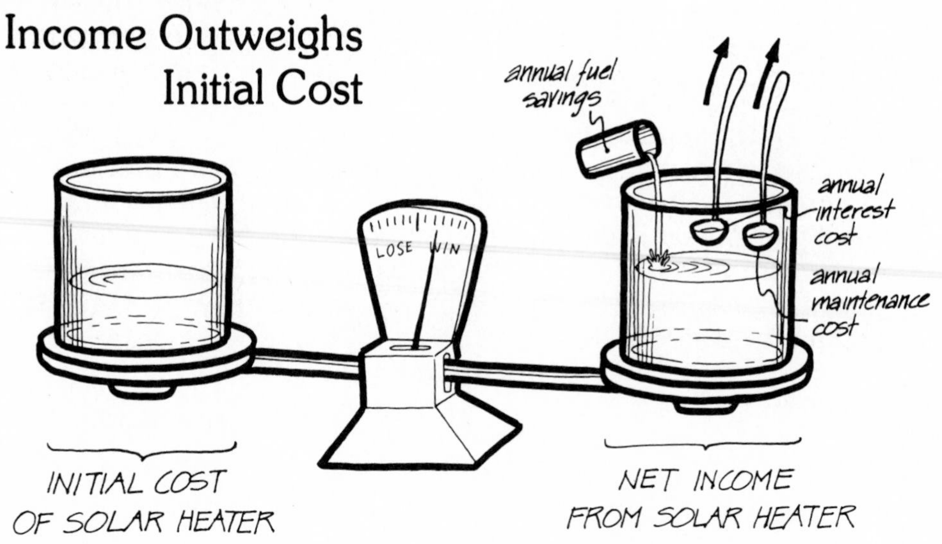

Trying to trade off these factors—solar heater cost, installation cost, maintenance cost, and fuel savings—is usually a very difficult process. And although cost alone is an important factor, it is not the only important one. Others are risk (will a solar heater last?), appearance (will it look ugly?), and convenience (will the house be cold in the morning?).

To understand how all these factors combine, think of conventional heating systems. In the United States, most houses are heated by natural gas, oil, or electric heaters. Electric heating is by far the most expensive "fuel"; the same amount of heat produced by electricity costs twice as much as heat produced by oil. Oil and natural gas heat cost about the same, although gas heat is still a little cheaper than oil heat. But in spite of the high operating cost of electricity, it is very cheap to install, requires virtually no maintenance, and is very convenient (allowing individual room thermostats, for example). Oil or gas heating, while yielding significant fuel savings compared with electric heating, have a higher initial cost, are difficult to install (there are lots of water pipes or air ducts), require more maintenance, probably won't last as long, and aren't as convenient. You might be surprised to know that despite the fuel savings of natural gas and oil, about half of the housing units being built today are electrically heated.

The decision as to which kind of solar heater to buy is similar to the decision as to which kind of conventional heater to buy. There are similar trade-offs between fuel savings, installation cost, maintenance cost, and the cost of the heating system itself. Just as no single conventional heating system is best, perhaps no single solar heater is best. Each type of system must find its own niche: a climate, a building style, or even a life-style that makes one system more suitable for you than another.

Glossary

absorption: The process by which solar radiation is converted to heat when it strikes a dark colored surface; also refers to radiation heat flow transferring to a surface.

active: A solar heating system that uses a pump (water-heating) to transfer solar heat to storage.

air-heating: A system using air as the heat-transfer medium in a solar heating system. Heat is usually stored in a rock-bed.

analogy: A device or function that is somehow equivalent to another device or function; for example, a gold mine is an economic analogy to a solar collector.

auxiliary: A furnace or other heating system used as a backup to a solar heater; for use when the solar heater can't provide all the heat required.

awnings: shading devices made of fabric or other material that block solar radiation; used for seasonal control

backup: An auxiliary heater used when the solar heating system can't provide all the heat required.

baseboard heaters: Heating devices in houses using either hot water or electricity that are installed along the baseboard of a room. They heat primarily by convection heat flow.

beadwall: A way to prevent nighttime heat-loss from a passive system by blowing styrofoam beads into the space between two glass layers on the south wall of a house. The beads are sucked out during the day.

biomass: Using the sun to grow plants such as trees or algae which are then harvested to produce energy.

boiling protection: Preventing damage in a solar hydronic system due to water inside boiling when the pump malfunctions.

caulking: Putting a plaster-like material in the cracks of a house or window to prevent infiltration heat loss.

changing phase: Changing from a solid to a liquid (melting) or from a liquid to a solid (solidifying). When a material changes phase it requires or gives up latent heat.

check valve: A valve that lets water flow through it in only one direction; analogous to the controller of an active solar heating system.

collector: The part of a solar heating system that captures the sunlight. Collectors gain heat from the sun but lose some of it by radiation and convection.

collector efficiency: The fraction of incoming solar radiation or rainwater that gets stored. Efficiency is low when the collector is hot and high when it's cool.

collector temperature: The average temperature of the solar collector (it might be a little hotter near the outlet and a little cooler at the inlet); analogous to pipe depth.

condense: Changing the phase of a material from gaseous phase to the liquid phase; for example, steam condenses into water; the opposite of evaporate.

conduction: Heat flow through a material where the material itself doesn't move; a type of thermal resistance.

conductor: A material which allows heat to flow through it easily (such as a metal); the opposite of an insulator.

conservation: Using less heat in a building by reducing the building's heat losses. More insulation, weather-stripping, caulking and storm windows all conserve heat.

control function: Controlling when and where solar heat is added to the living space from storage.

controller: A device that controls when the pump or fan on a solar heating system should be turned on. It usually uses information from collector and storage temperature sensors.

convection: Heat flow from a surface into the surrounding gas or liquid (usually air or water); a type of thermal resistance.

conventional heater: A heating system other than solar; for example, gas-fired furnaces, baseboard heaters, oil furnaces.

corrosion: When metals disintegrate, usually because of chemical action by oxygen (e.g., rusting) or other substances. Usually water-heating systems have more corrosion problems than air-heating systems.

crushed rock: Heat storage material used in rock-bed air-collector systems. Rocks are usually about the size of an egg.

daily control: Controlling solar heat over a day's period: not too much heat in late afternoon but heat stored for use at night; for example, a masonry wall.

depth: The level of liquid in a container; analogous to temperature.

depth difference: One level subtracted from another level; difference between two levels or depths; analogous to temperature difference.

diode function: The method by which solar heating system allows heat to go to storage when it's available but prevents heat loss from storage at night; for example, a controller in an active system provides a diode function.

direct-gain: A passive system where the sun heats the floor or walls of the house directly through south-facing windows.

direct use: Using the sun's radiation directly as in heating and generating electricity.

distribution: Getting heat to the living space. In a passive house, distribution to north-facing rooms is often difficult.

double-glazing: Using two layers of glass on windows or solar collectors to reduce heat loss. The still air in the space between the windows acts as a good insulator.

downhill: Direction of heat flow (from a higher temperature to a lower one); direction of volume flow (from a higher depth to a lower one).

drain-down: Draining a solar hydronic system at night to prevent water from freezing in its collectors.

drumwall: A passive system which uses water-filled drums for heat storage and a hinged, insulating wall to prevent nighttime heat loss.

electric furnace: A heater where forced air is blown past electrically-heated coils. The air is heated by convection from the coils' surface.

electromagnetic radiation: A class of radiation whose different "pitches" include solar radiation, radiation heat flow, x-rays, light, microwaves, T.V. waves and radio waves.

evacuated tube collector: A collector with a vacuum jacket and reflectors that reduces heat losses and improves efficiency at very high temperatures.

evaporate: Changing the phase of a material from liquid phase to gaseous phase; for example, water evaporates to form water vapor (steam); the opposite of condense

fan: A device which blows air in an air-heating solar system or in a forced-air heating system. A fan moves heat by transport heat flow.

fiberglass batting: An insulator made from loosely-packed glass fibers, coming in thick layers which can be stapled to a wall or layed between ceiling rafters. Heat flows by conduction through batting since the fibers prevent convection.

fin: A thin sheet of metal in a heat exchanger that conducts heat to the surrounding gas or liquid.

first cost: The amount you pay for a solar heater when you first buy it; sometimes called the initial cost.

flap valve: A check valve that uses a flap to block the liquid flow in one direction, but allow flow in the other direction.

flat-plate collector: A collector which remains fixed in position; the collecting part of a flat-plate solar heating system.

fluid resistance: Impediment or resistance to fluid flow. A high fluid resistance means fluid flows with difficulty, a low fluid resistance means fluid flows easily.

forced-air: Heating a house using heated air forced into the rooms by fans.

forced convection: Convection by a gas or liquid forced over a surface; for example, a pump forcing water through a pipe or a fan blowing air over a brick—in contrast to natural convection where an object's temperature alone causes flow past its surface.

fossil fuels: Fuels derived from the fossils of plants that died millions of years ago; oil, natural gas and coal are all fossil fuels.

freeze protection: Protecting a solar hydronic system at night to prevent water from freezing in its collector.

fuel savings: The money you make each year on solar heater because it's eliminating some of the fuel costs you would otherwise have.

glazing: Transparent cover (or covers) that prevent heat losses from a solar collector. Glazing is usually made of plastic film or glass.

glycol: An antifreeze additive to water.

ground reflection: Solar radiation reflected from the ground (e.g., by snow) into a solar collector; more heat can be captured by a collector with ground reflection.

greenhouse: A passive system which separates the collector and storage from the living space by means of a greenhouse buffer.

greenhouse effect: The property of glass where solar radiation passes through but radiation heat flow is blocked. In greenhouses, sunlight enters but heat is trapped by the glass.

heat: A quantity indicating how much energy a substance contains; analogous to volume.

heat exchanger: A device which is designed to transfer heat easily. Heat transfer can be between a liquid and gas or between liquid and liquid; analogous to the tap on the rainwater collecting system.

heat flow: The movement of heat from one place to another due to temperature differences; analogous to volume flow.

heat-flow path: Path which heat takes as it moves by conduction through a material from a high temperature to a lower one. Length of flow path is how far the heat travels, area of the flow path is how much material the heat flows through at each point along the path.

heat losses: Heat flow from an object that is lost; for example, heat flowing from a house on a cold day are house heat losses.

heat storage: Keeping heat collected at one time for use later on; analogous to volume storage. Heat is stored by materials when they get hotter or when they're melted.

heat store: Part of a solar heating system that stores heat.

heat transfer medium: A gas or liquid that moves heat by either convection or transport heat flow.

high-pitched: In sound, radiation refers to high notes that come from instruments like violins and flutes. In electromagnetic radiation it refers to solar radiation (and light and x-rays, as well).

hot-water heating: Solar energy used to heat water for showers, laundries, cooking, dishwashing and so forth.

house heating: Solar energy used to heat a house. (*See* space-heating.)

hydronic: A solar heating system that heats water in its collector and stores heat in a water tank; a water-based heating system.

indirect use: Using the effects of the sun's radiation for energy. Fossil fuels, biomass wind power and ocean currents are examples.

infiltration: Heat loss from a house due to cold air leaking in cracks and warm air leaking out through cracks or chimneys. Some infiltration is needed or a house feels stuffy.

initial cost: The amount you pay for a solar heater when you first buy it; sometimes called the first cost.

inlet: Where the air or water enters a solar collector or other device.

installation: The cost of getting a solar system from the crates its delivered in to a working solar heater. Installation costs are a part of the first cost.

insulated-plate collector: An unglazed solar collector with an insulated metal plate but no means for extracting solar heat.

insulation: General measure of thermal resistance. Insulation resists the flow of heat.

insulator: A material that prevents heat from flowing easily by conduction. High insulation corresponds to high thermal resistance; opposite to a conductor.

interest: Interest is the cost of money.

latent heat: Heat stored by melting and solidifying a substance, such as wax. Phase change materials store heat latently.

latitude: How far a point on the earth's surface is from the equator. A solar collector near the equator (low latitude) usually has a smaller tilt angle than one nearer to the poles (high latitude).

leafy plants: Deciduous plants which block the sun in summer but allow solar heating in winter; used for seasonal control of solar radiation.

living space: The space in a house that people live in.

liquid-to-air heat exchanger: A device that transfers heat from liquid to a gas; for example, a car radiator or a baseboard heater.

liquid-to-liquid heat exchanger: A device that transfers heat from a liquid to another liquid; for example, a coil in a heat storage tank.

low-pitched: In sound radiation it refers to low notes that come from instruments like basses and tubas. In electromagnetic radiation it refers to radiation heat flow (and radio, T.V. and microwaves as well).

maintenance: Keeping your solar heater working will cost you either time or money. Maintenance costs reduce the income you get each year from your solar heater.

masonry wall: A passive solar house that lets solar radiation heat a glass-covered masonry wall. Heat is conducted into the house through the wall.

mass-production: Making products in a factory where less human labor is needed. The result is that most things can be produced cheaper in a factory.

melting: Changing phase from a solid to a liquid. Most phase change materials (like wax) require extra (or latent) heat melt.

modular: A solar system that's made in individual units or modules, each module having the capability of an entire system; for example, a thermic diode is a modular solar heating system.

movable insulation: Moving insulation on a passive solar heating system in a way to prevent heat loss at night; for example, "Drum-wall" and "Skytherm".

natural convection: Convection caused solely by an object's temperature. Gas or liquid near a hot surface rises, near a cold surface falls; differs from forced convection where the gas or liquid is forced over the surface.

new income: The income you actually get from a solar heater each year—annual fuel savings less maintenance and interest.

nighttime loss: Heat loss from a solar collector at night. A controller of an active system turns off the pump or fan at night to prevent nighttime heat loss. Passive systems use movable insulation and other means.

oil valve: A check valve in a thermic diode uses oil in a chamber to block flow in one direction but allow flow in the other direction.

outlet: Where the air or water leaves a solar colector or other device.

overcast day: A day when solar radiation is blocked by clouds or haze. A cloudy day may have only a fifth the solar radiation of a clear day.

over-heat: The tendency for some passive solar heating systems to get too hot in the late afternoon.

parallel configuration: Where a solar heater and auxilliary heater combine to heat a house; in contrast to a series or preheater arrangement.

passive: A solar heating system that doesn't use fans or pumps to transfer solar heat to storage.

phase-change materials: Materials that store heat latently by melting and solidifying, that is, by changing phase.

photovoltaic: Devices which convert sunlight directly to electricity.

pre-heater: A way to use a solar heating system where solar heat warms the air or water and then an auxilliary heater or furnace finishes the heating by adding additional heat to the air or water.

pump: A device which forces water through an active solar heating system; analogous to the fat-connecting tube on a rainwater collector system.

pyramidal optical condenser: A solar heating system that has relfectors mounted in the roof peak of the house.
radiation: Heat flow from a surface by electromagnetic waves; no surrounding liquid or gas is necessary (can occur in a vacuum); a type of thermal resistance.
radiator: A device in a home or car to let heat flow easily from heated water into the air. In a car ir cools the car's engine, in a home it warms a room. Radiators usually transfer more heat by convection than by radiation heat flow.
rainwater collector: A tray used to catch rain waater; analogous to a solar collector.
rainwater equivalent: A part of or function on a rainwater collector system that is analogous to a solar collector system.
reflection: The process by which solar radiation, on striking a light-colored surface, is not absorbed—it reflects away; also refers to radiation heat flow on striing a shiny surface.
reflector: A shiny surface that reflects sunlight into a collector to increase the solar radiation it absorbs.
regulated hot water: Hot water whose temperature is held more or less fixed.
reliable: A solar heating system that's rugged and simple and expected to operate a long time without need of repair.
retrofit: Installing a solar system on an existing house rather than on one that's just being built.
return hose: A hose in a thermic diode that lets cool storage water return to the collector.
reverse thermosyphon: A way in which a passive thermosyphon system can lose heat at night by the air or water circulating in the reverse direction.
rock-bed: The storage medium for an active air-heating collector system. Rocks are crushed and held in an air tight container with appropriate ducts.
seasonal control: Controlling solar heat through dthe various seasons; for example, awnings block the summer sun but allow solar heat in winter.
selective coating: A coating on the absorber plate of a solar collector that absorbs solar radiation but prevents radiation heat flow.
selective surface: A solar collector absorbing surface with a selective coating. Radiation heat loss from the surface is reduced.
series configuration: A solar heater used as a pre-heater to an auxiliary heater.
set temperature: The temperature a thermostat is set to.

silicon: A material that's used to make photovoltaic solar cells that convert sunlight to electricity.

silvered surfaces: Shiny, mirror-like surfaces which reflect both radiation heat flow and solar radiation.

sky temperature: The temperature of the sky. Usually it's cooler than the outdoor temperature, especially when there are no clouds and where it's very dry (as the desert).

Skytherm: A passive system with rooftop water bags storing heat and insulating slabs that slide over the water bags at night to prevent heat loss.

solar air-conditioning: Using solar heat to cool a building. Often evacuated tube collectors with low heat loss must be used.

solar cells: Photovoltaic devices which convert solar radiation directly to electricity; used in space satellities, watches and calculators.

solar radiation: A type of electromagnetic radiation coming from the sun that heats objects when it's absorbed. Solar radiation is reflected by light-colored surfaces and absorbed by dark-colored ones. It goes through transparent materials like glass and plastic.

solidifying: Changing phase from a liquid to a solid. Most phase-change materials (like wax) give up extra or latent heat on solidifying.

sound radiation: A class of radiation that lets us heat. Sound radiation is divided into high pitches (violin and flute sounds) and low pitches (base and tuba sounds).

south-facing: Toward the equator; the best direction to point a solar collector in the northern hemisphere. In the southern hemisphere, for example, Australia, the best collector orientation is north-facing.

space heating: Heating of a house's living space; in contrast to hot water heating and swimming pool heating.

stagnation depth: The depth a rainwater collector gets when no rain water is being saved. The stagnation depth is higher when more rain is caught; analogous to stagnation temperature.

stagnation temperature: The temperature a solar collector gets when no heat is being saved. It's higher when more sunlight is falling on the collector, and when the collector is glazed; analogous to stagnation depth.

still air: Air which isn't moving. Heat loss from a surface in still air is by natural convection and is much less than if the air is moving.

storage material: Material that stores heat in a solar heating system. Storage materials are rock-beds, water tanks, concrete walls and water-filled drums.

storage temperature: The average temperature of the heat stored (it might be a little hotter at the top and a little cooler at the bottom because of stratification); analogous to tank depth.

stratification: The tendency for heated air or water to stay at the top of a storage container.

sun louvers: A passive solar space-heating system where sunlight is reflected by louvers into ceiling-mounted phase-change material. The phase-change material provides nighttime heat as it solidifies.

sun window: A system which uses a dark fluid to block solar radiation when it's not needed. The dark fluid is pumped through passages within the collector.

surface area: The outside surface of an object that loses heat by convection or radiation. The bigger the surface area, the more heat lost.

swimming pool heating: Solar energy used to heat swimming pool water. The pool itself stores the heat.

tap: An outlet on the side of a rainwater storage tank from which to draw stored water; analogous to the heat exchanger of a solar heating system.

tax credits: Many states and the Federal government give tax relief to people who use solar equipment. Tax credits reduce the solar heater's first cost.

temperature: The hotness of an object measured with a thermometer; analogous to depth.

temperature difference: One temperature substracted from another temperature; analogous to depth difference.

temperature rise: The temperature difference between inlet and outlet of a solar collector; the temperature of a heat storage tank above the temperature to which it delivers heat.

temperature sensor: An electronic device that measures the temperature of collector or storage; used to signal a controller when to turn on a pump or fan.

temperature swing: The typical change in temperature between early morning and late afternoon in a passive solar heating system.

thermal contact: Two objects that can transfer heat easily from one to the other, usually by conduction.

thermal resistance: Resistance or impediment to heat flow. A high thermal resistance means heat flows with difficulty, a low thermal resistance means heat flows easily; measure of how well-insulated something is.

thermic diode: A passive solar heating module that works on the thermosyphon principle. It has an inherent diode function to prevent heat loss at night.

thermodynamic cycle: A method of using heat to give a cooling effect by evaporating (boiling) a liquid and then condensing (liquefying) it again.

thermostat: A device used to control when a furnace turns on and off depending on the temperature of the room it's in. A thermostat tries to hold the room's temperature at a fixed point.

thermosyphon: Heat transferred by a circulation process where heated water or air rises and is replaced by cooler water or air. Trombe walls and thermic diodes use thermosyphon action.

tilt angle: The angle a solar collector makes with the horizontal. Besides facing south, collectors have a high angle for house heating, medium for hot water heating, and low for swimming pool heating.

tracking system: Solar collectors which follow the sun. Also tracking mirrors follow the sun, keeping the sun's reflection focused on a stationary point.

transmission: The process by which solar radiation goes right through transparent materials such as glass, plastic and air.

transparent: A material through which light can pass such as glass or plastic film.

transport: Heat flow by transporting or moving a substance which is hot. Often the substance moved is a gas or liquid (air or water); a type of thermal resistance.

trombe wall: A passive solar house using a glass-covered masonry wall with air holes at top and bottom of the wall. Heat is transferred to the house by conduction through the wall and thermosyphon through the air holes.

unglazed: A collector with no glazing; convection heat losses are high without glazing since the absorbing surface is exposed to the wind.

useful heat: The heat stored above room temperature in a house heating application. Heat stored below room temperature isn't useful because it can't warm the room (except as a pre-heat for a furnace); analogous to useful volume.

useful volume: The volume stored in a tank above the outlet

tap; analogous to useful heat.

vacuum jacket: A vacuum around an evacuated tube collector that prevents convection heat loss.

vapor: The gaseous phase of a material; for example, steam is the gaseous phase of water.

volume: The amount of space that a liquid occupies; analogous to heat.

volume flow: The movement of volume from one place to another; analogous to heat flow.

water-heating: Using water as the heat transfer medium in a solar heating system; a hydronic system.

water main: The pipe that brings cold water into a house from a residential water system.

weatherstripping: Putting felt or foam rubber around doors and windows to prevent heat loss by infiltration.

wind power: Using the wind to produce power such as to pump water, grind grain or generate electricity. The wind is indirectly caused by solar energy.

Index